Chemicals

What you Need to Know

RONALD ELLROY NEY, JR., Ph.D.

Published by
Chatelaine Press
6454 Honey Tree Court
Burke VA 22015.

Written by Ronald E. Ney, Jr.

Edited by A.G. Lantzman-Forbes

Second Edition

Fully revised

CIP96-4517 *ISBN 1-57420-053-4*

Ney, Ronald E., 1936
1. Toxicology–popular works. 2. Household supplies–Toxicology. 3. Chemicals–. I. Title.
RA1213.N44 1996
363.17'91–dc20

Cover by Kathleen Myers

DISCLAIMER

Chatelaine Press and the author do not recommend and do not suggest discontinuing the use of any prescribed medications; call your doctor if you have any questions.

Pesticide products and other potentially *hazardous* or *allergenic* products contain safety instructions. Be sure to follow all safety instructions on the laboratoryels or package inserts.

Use of protective clothing (i.e. gloves, long sleeves, eye protectors) is a good habit to develop when using *pesticide*s and strong cleaning products.

Careful consideration has been given to each entry to give useful information to the user. However, neither Chatelaine Press nor the author shall be held responsible or liable for any damage resulting in connection with use of the data provided in this book. The purpose of this work is only to inform and no recommendations are intended.

About the Author

Ronald E. Ney, Jr. Ph.D. is a physical scientist, with a master's degree in chemistry and a doctorate in physical science. He was formerly a member of the US Environmental Protection Agency where his positions included Chief Chemist for the Hazard Evaluation Division, Chemist Advisor in the Office of Solid Waste, Liaison to EPA's Office of Research and Development, and Section Chief in the *Pesticide* Registration Division His vitae includes positions as Research Director and President of an Environmental Research Consulting Service. He also served as Supervisory Chemist in the Department of Agriculture's *Pesticide* Registration Division and as a Chemistry Laboratory group leader for the Food and Drug Administration's studies on *pesticide* residue metholology.

Currently Dr. Ney is President and Chief Investigator for a consulting firm specializing in chemicals in realty. He is also an adjunct professor at Northern Virginia Community College where he teaches courses in Environmental Issues and Topics in Environmental Risk in Real Estate Transactions.

He is a Registered Environmental Assessor, Certified Contract Officer (EPA) and Registered Environmental Professional. He has written two previous chemistry books: *Where Did That*

Chemical Go? and *Fate and Transport of Organic Chemicals in the Environment.*

He is currently writing a high school chemistry text book .

INTRODUCTION

Chemicals is intended as a guide to help consumers make informed choices about their purchases and understand some of the hazards that accompany the benefits of the myriad of products available in today's market.

This book starts with the assumption that you are aware of the benefits of the products you purchase and thus dwells on the negative to inform you of potential adverse effects.

Almost everything is toxic at some level. It is not possible to define dose response levels in the context of this work. Body weight, general physical condition, family history and allergic tendencies will affect the reaction threshold for each individual user.

However, if you find that you or others in the family have a reaction to products you use, this book may help you to identify and eliminate those specific chemicals that cause your adverse reactions. To do this you will have to consider the individual product ingredients.

Chemicals is not all inclusive. Not every chemical, every chemical name, every use, or every hazard can be listed. This directory contains a list of over 600 chemicals and biologicals along with their uses and possible hazards. There is space after each entry to enter your personal notes. *Chemicals* will help you to be aware of the effects of chemicals in common use.

HOW TO USE *CHEMICALS*. Look at the ingredients on a product label and then look for that ingredient in your guide. (In addition to chemicals in manufactured products, we have included plants and plant products that may grow in residential areas or may be purchased in fruit and vegetable markets.) The listing will briefly inform you of some of its uses and hazards. Italicized words are further defined in the glossary. With this knowledge, you will be in a position to make an informed purchase and choose the chemicals to which are willing to be exposed. This guide will enable you to have control over the hazardous chemicals in your home and environment"

Following each entry you will find space to write your own personal notes.

ABOUT THE SCALE

Chemicals in this books are variously identified as being "possible", "suspected" "experimental" or "confirmed" *carcinogens*, *teratogens*, or *allergens*. Possible means there is some literature to support the belief, but no research. Suspected is a bit stronger in that there is some research but no conclusive results. Experimental means laboratory results have been obtained, but continuing research is required. Confirmed means that laboratory results have been widely replicated and confirmed by the scientific community.

A FINAL WORD

One important piece of information which I want to share with you, is the following: many will say exposure to chemicals and/or biologicals in consumer products or in the environment is so small there is little chance of risk. While this may be true, safety cannot be judged on one chemical or one biological alone. Humans and other animals are a mixture of chemicals and biologicals, and we take in hundreds of different chemicals and biologicals a year. Just how safe are the chemicals and biologicals when the *aggregate*, *synergistic*, *antagonistic*, *co-metabolism*, and *co-biometabolism* effects are never mentioned or even studied to any extent, if at all and, they are not used in risk *assessments*? In other words, the total picture is never known or considered for human and environmental safety.

Let *Chemicals* help you in your daily buying decisions.

a

ACARBEN—*see* chlorobensilate.

ACCELERATE—also called disodium-3,6-endoxohexahydrophthalate—is a *pesticide* (*herbicide*) which is an eye, skin and mucous membrane *irritant*. It is also *toxic* by ingestion and skin contact.

ACETALDEHYDE—also called acetic aldehyde—is used in flavors and perfumes. It is a mucous membrane *irritant* and, in large doses, it may even cause death. It is also a suspected *carcinogen*.

ACETIC ALDEHYDE—*see* acetaldehyde.

ACETONE is both a *pesticide* and a solvent. It is an eye *irritant* and, if ingested, can cause kidney damage. In addition, it is *flammable*.

ACETYLACETONE is used as both a *fungicide* and *insecticide*. It is also used as a gasoline additive, a drier for printer's ink and in varnishes. It is an *irritant* to eyes, skin and mucous membranes.

ACETYLEUGENOL—also called eugenol acetate—is a flavoring agent and perfume ingredient. It is a skin *irritant* which is toxic by ingestion.

ACETYLSALICYLIC ACID—*see* aspirin.

ACROLEIN—also called acrylaldehyde—is used in the manufacture of perfumes and pharmaceuticals. It is an eye *irritant*. This chemical is metabolized in the body to form the metabolite glycidaldehyde, a suspected *carcinogen*.

ACRYLALDEHYDE—*see* acrolein.

ACRYLAMIDE—also called propenamide—is a waterproofing agent and soil stabilizer. It has been shown to be *carcinogenic* to laboratory animals and is a suspected human *carcinogen*. Acrylamide is also a central nervous system depressant.

ACRYLIC ACID—also called vinylformic acid—is used in water purification systems; in medical and dental materials; and in coatings on paper and cloth.. It is an eye, nose and skin *irritant*.

ACRYLONITRILE—also called 2-propenetrile; vinyl cyanide—is used in a variety of applications. These include acrylic fibers; plastics; adhesives and *pesticide* (fumigant for stored grain). It is a human *carcinogen* and may form poisonous cyanide gas.

ACTIVOL—*see* para-aminophenol.

AFLATOXINS comprise a group of highly toxic polynuclear mycotoxins (mold). These are naturally occurring contaminants in fruits, vegetables, cereal grains and peanuts. Since aflatoxins are a mold, they can be eliminated through proper storage and elimination of stale, moldy, and out-of-date products.

ALAR—also called Diaminozide—is a *pesticide* and is also used to regulate plant growth. It is a *carcinogen.*

ALBRIN—also called toxalbumin—is a toxic *lectin* from the seed of a common vine (*Abrus precatorius*) found in central and southern Florida. One thoroughly chewed seed can be fatal.

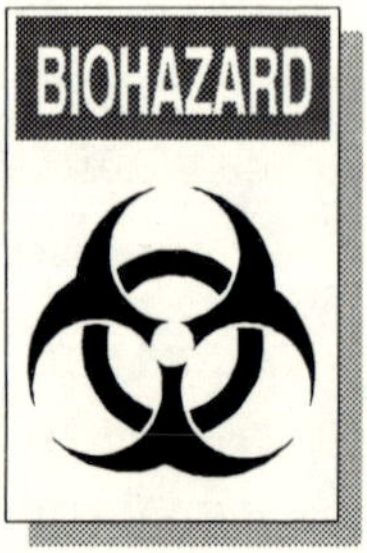

ALCOHOL—*see* ethyl alcohol.

ALDICARB—also called Temik—is a *pesticide* (*nematocide* and *insecticide*) which is poisonous by ingestion and skin contact.

para-ALLYLANISOLE—also Tarragon—is a naturally occurring oil derived from anise and basil. It is used in foods, liqueurs and perfumes. It is a skin *irritant* and possible *carcinogen.*

ALLYL ISOTHIOCYANATE—also *volatile* oil of mustard or Redskin—is a medical *counterirritant* (mustard plaster), a food condiment and popular flavorant. In small concentrations, it is the main ingredient in common table mustard, while in extremely high concentrations it is the major component of mustard gas. It is an allergen. It is a suspected *carcinogen* and is poisonous by skin contact and ingestion. *Teratogenic* and reproductive effects have been shown in laboratory, and it breaks down to poisonous cyanide gas.

ALMOND

(As readers of mystery novels know, cyanide poison has an almond odor.)

BITTER ALMOND—the rape seed of *Prunus amygdalus Stokes var amara Focke*—is used in *perfumes*, and the manufacturing of liqueurs. Cyanide poisoning has occurred from ingestion.

SWEET ALMOND—the rape seed of *Prunus amygdalus Stokes var dulcis (D.C.) Baill*—is used in *perfumes* and as a confection. Sweet almond is also used in almond milk.

ALUMINA—*see* aluminum oxide.

ALUMINUM is used as a *pesticide*. It is also used in buildings and construction and for electrical uses. Dust particles may cause pulmonary *fibrosis*. Fine powder may explode.

ALUMINUM CHLOROHYDRATE is a common component of deodorants. It is also used for water purification and in the treatment of sewage and some plant discharges.

ALUMINUM OXIDE—also called alumina—is a filler for paints and varnishes. It is also used in dental cements and ceramics. It is a questionable *carcinogen* in experimental animals. Inhalation of fine particles may cause lung damage.

ALUMINUM SILICATE is a component of dental cements, enamels, ceramics, and paint fillers.

ALUMINUM SODIUM SULFATE—also called soda alum—is a *food additive* and medical *astringent*. It may cause *dermatitis* or skin irritations. If burned, it breaks down to toxic fumes.

para-AMINOBENZOIC ACID—also called Vitamin H—is used as a topical sunscreen (ultraviolet screen). Laboratory animals have shown reproductive effects. Ingestion can result in nausea or vomiting. It can cause skin rash.

4-AMINOBIPHENYL—*see* para-biphenylamine.

4-AMINO-1-HYDROXYBENZENE—*see* paraaminobenzoic.

para-AMINOPHENOL (PAP)—also called 4-amino-1-hydroxybenzene; Activol—used in the preparation of pharmaceuticals. It is used in hair dyes. *Teratogenic* effects have been noted in rats.

AMMONIA is a fertilizer and a refrigerant. It is also a common household cleaning product. Ammonia is poisonous by inhalation and ingestion as well as an eye, skin and mucous membrane *irritant*. Mixing ammonia products with chlorine products produces a deadly gas.

AMMONIUM LAURYL SULFATE is a *pesticide* used in shampoos.

AMMONIUM XYLENESULFONATE is used in hair shampoo.

AMPHETAMINE—*see* benzedrine.

ANILINE—also called aniuline oil—is a *pesticide* used in the manufacture of dyes, medicinals, perfumes, varnishes, and shoe black. It is a suspected *carcinogen* and an eye and skin *irritant*

ANIULINE OIL—*see* aniline.

ANTRACINE 12—*see* BHA.

APPLES—*see* potatoes.

ARECA CATECHU—*see* betel nut.

ARECOLINE is a veterinary medicine used to kill tape worms in dogs and poultry, as a *cathartic* in horses, and a *ruminatoric* in cattle. It may cause tremors, and is a *mutagen*. It is a questionable *carcinogen* in laboratory animals.

ARSENIC is a highly toxic *pesticide*. It is a *carcinogen* and *mutagen* which is *teratogenic* in laboratory animals. Some arsenic compounds are listed below and they may have these hazards.ARSENIC ACID—also called orthoarsenic acid—is a *pesticide* and is also used in the manufacture of glass and as a wood treatment.

ARSENIC DISULFIDE is a *pesticide* and is used as a colorant in paint pigment.

ARSENIC PENTOXIDE—also arsenic acid—is a *pesticide* and is used as a coloring agent in dyeing and printing and in colored glass.

ARSENIC SULFIDE is a *pesticide* and is used in paints.

ARSENIC TRIOXIDE—also called white arsenic—is a *pesticide*. It is used in pigment, ceramic enamels, and *wood preservative*.

CACODYLIC ACID—also called dimethylarsenic acid—is a *pesticide*. It is an eye and skin *irritant*, questionable *carcinogen*, and an experimental *teratogen*.

COPPER ACETOARSENATE—also called paris green—is a *pesticide*. It is used in *wood preservatives* and in antifouling paints.

COPPER ARSENATE is a *pesticide*.

COPPER ARSENITE is a *pesticide* and is used in *wood preservatives* and as a coloring agent in pigments.

DISODIUM METHANEARSONATE (DSMA) is a *pesticide*.

LEAD ARSENATE is a *pesticide* which is used in waterproof paints and antifouling paints.

MERCURIC ARSENITE is a *pesticide* and is used in both waterproof and antifouling paints.

MONOAMMONIUM METHANEARSONATE is a *pesticide*.

MONOSODIUM ACID METHANEARSONATE is a *pesticide*.

POTASSIUM ARSENATE is a *pesticide* and is used in fly paper.

SODIUM ARSENATE is a *pesticide* and is used in dyeing and printing.

SODIUM ARSENITE is a *pesticide* and is used as an antiseptic and in dyeing.

ZINC ARSENATE is a *pesticide* and is used in *wood preservative*.

ZINC ARSENITE is a *pesticide*.

ASBESTOS is a naturally occurring mineral. It is used in fireproofing material; paints, cement, brake linings, insulation, and shingles. It is a confirmed *carcinogen* which causes lung cancer.

ASPARTAME—also called Nutrasweet—is a sweetener used as a sugar substitute. It may cause allergic *dermatitis*. It has caused brain tumors in animal laboratory tests.

ASPHALT—also called mineral pitch—is a petroleum derivative commonly used as road tar and on roofing tar. It is an *irritant* and a suspected *carcinogen*.

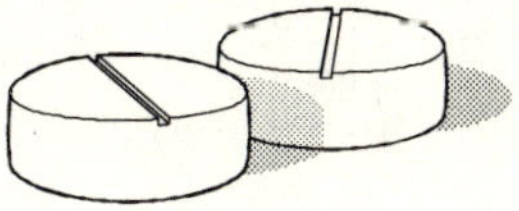

ASPIRIN—also called acetylsalicylic acid—is a white, crystalline compound, commonly used in tablet form to relieve pain and reduce fever and inflammation. It may cause menstrual cycle changes, nausea, or liver damage and is an allergen. It is contraindicated in conjunction with anticlotting drugs. It is a possible human *teratogen*. In children, taking aspirin for flu symptoms is sometimes associated with Reye Syndrome. Its use can be fatal.

AUREOMYCIN—*see* chlortetracycline.

AVADEX—*see* diallate.

AZINPHOS METHYL—also called methyl guthion—is a *pesticide* (insecticide) and questionable *carcinogen* in laboratory animals.

AZOBENZENE is a *pesticide* (*acaricide*). It is a questionable *carcinogen*.

b

BALSAM—also called balsam of Peru and balsam of Tolu—is a naturally occurring aromatic plant resin used in *perfumes* and chocolate flavoring. It is used medically in skin lotions and cough preparations; to treat skin ulcers and as a *scabicide*. In veterinary practice balsam is used as a *miticide* and for healing wounds. It is an allergen.

BANANA OIL—*see* isoamyl acetate.

BARIUM SULFATE is used in the manufacture of cellophane, linoleum, and paper and as a color pigment for paper and wallpaper. It is also a medical diagnostic aid. It is a suspected *carcinogen.*

BASIL OIL—also oil of basil—is derived from a naturally occurring plant substance (herb) and is toxic by ingestion. It is an eye and skin *irritant*.

BAY OIL—also oil of bay—is a naturally occurring herb. It is toxic by ingestion.

BEESWAX is derived from natural honeycomb and is used in the manufacture of wax paper, as a food, and as a pharmaceutical aid. It is an *allergen*.

BELLADONNA—also called death's head and poison black cherry—is a naturally occurring *alkaloid* in perennial plants such as belladonna and jimsonweed. Taken internally, it is a medical *anticholinergic* and deadly poison. It may cause contact *dermatitis.*

BENOMYL is a *pesticide* (*fungicide*) also used as an oxidizer in sewage treatment. It is a skin *irritant*, *carcinogen; mutagen*; and *teratogen.*

BENTONITE is an absorbant aluminum silicate clay formed from volcanic ash and used as an emulsifier for oils and plaster and a pharmaceutical aid. It is a suspected *carcinogen.*

BENZALDEHYDE—also benzoic aldehyde; malachite green—is a *pesticide* used in *perfumes*, dyes, solvents, and flavors. It is an allergen and is toxic by ingestion.

1,2-BENZANTHRACENE—also benz [alpha] anthracene; tetraphene—is used in oils, waxes, food, and drugs. When burned, it gives off a toxic smoke. It is a *carcinogen.*

BENZ [ALPHA] ANTHRACENE—*see* 1,2-benzanthracene.

BENZEDRINE—also amphetamine—is a medical central nervous system stimulant for humans and animals. It is used as an *anorexic* and can raise blood pressure. It is also a deadly human poison.

BENZENE—also benzol—is a *pesticide*. It is used in solvent, paint remover, and gasoline. It is a central nervous system depressant and *carcinogen*. There is both *fetotoxin* and *teratogenic* evidence.

BENZENE HEXACHLORIDE—*see* lindane.

BENZIDINE is used in the manufacture of dyes. It is a *reagent* for hydrogen peroxide. It damages blood and bone marrow and is a *carcinogen* which causes bladder tumors.

BENZOIC ACID is a *pesticide*. It is used as a food preservative, in juices, in the manufacture of dyes and curing of tobacco. It is also used as a topical antifungal with salicylic acid. It is an eye, skin and mucous membrane *irritant*.

BENZOIC ALDEHYDE—*see* benzaldehyde.

BENZOL—*see* benzene.

BENZYL ACETATE is used in perfumes. It is an experimental *carcinogen* in laboratory tests. It is an eye, skin and respiratory tract *irritant*.

BENZYL ALCOHOL is a *pesticide*. It is used in perfume, as a pharmaceutical aid, as a gelatin solvent, and in shellac. It is an eye and skin *irritant* and is toxic both by inhalation and skin contact.

BENZYL CARBINOL—*see* 2-phenylethanol.

BENZYL CHLORIDE is used in the manufacture of perfumes, pharmaceuticals, and dyes. It is a suspected *carcinogen*. It is an eye, skin and mucous membrane *irritant*; *central nervous system depressant*; and is poisonous by inhalation.

BENZYL CINNAMATE is used in artificial flavors, and in perfumes. It is an allergen and a skin *irritant*.

BENZYL ETHER is a perfume solvent. It is an eye and skin *irritant*.

BENZYL PENICILLIN SODIUM—also called penicillin G. sodium salt—is a medical and veterinary *antibacterial*. It is a possible *carcinogen*. It is toxic.

BENZYL SALICYLATE is used in perfumes and in sunscreens. It is toxic by ingestion.

BERBERIN—*see* berberine.

BERBERINE—also called berberin—is an *antibacterial,* antimalarial and antipyretic. It is poisonous and can cause death by central paralysis.

BERGAMOT OIL (if rectified it also contains alinalyl acetate, 1-linalool, d-limonene, dipentene, bergaptene) is naturally occurring. It promotes sun tanning and is *antisporiatic*. It is a skin *irritant* and an allergen and is toxic by ingestion.

BETEL NUT—also called areca catechu; pinang—is from a naturally occurring tropical plant and is used as a *counterirritant*. It is a suspected *carcinogen* in laboratory animals.

BHA (Butylated hydroxyanisole)—also called antracine 12—is used as an *antioxidant* in food, natural fats and oils. It is an experimental *carcinogen* in laboratory tests.

BHT—also called butylated hydroxytoluene—is used in soaps. *See also* cresols.

para-BIPHENYLAMINE—also called 4-aminobiphenyl—was formerly used as a rubber *antioxidant*. It is a *carcinogen* and has caused tumors in the mammary gland, intestinc, and bladder.

BIRCH TAR OIL is a naturally occurring substance derived from the Birch tree. It is used to preserve leather and wood. It is also used in dermatology and to treat skin diseases in veterinarian medicine. It is an eye and mucous membrane *irritant* and an allergen.

BITTER ALMOND—the rape seed of *Prunus amygdalus Stokes var amara Focke*—is used in perfumes and the manufacturing of liqueurs. Cyanide poisoning has occurred from its ingestion.

BLACK pepper naturally contains safrole and piperdine.

SAFROLE is a *pesticide* and is a *carcinogen* in laboratory animals. It is no longer on the market.

PIPERDINE is a *pesticide* (*insecticide*). It is in brandy. It is toxic by ingestion and is a skin *irritant.*

BLEOMYCINS are part of a group of related gylcopeotide antibiotics. The bleomycins are *antineoplastics* and are suspected *carcinogens.*

BORIC ACID is a *pesticide*, is used as a food preservative, in perfumes, for flavor, and in mouthwash. It is a human *astringent* and is an antiseptic. It is a veterinary *antibacterial* and is an antifungal. It is toxic by ingestion, an eye *irritant*, and can cause nausea, vomiting, diarrhea, convulsions, and coma. It has caused death in both adults and infants.

BORNYL ACETATE is used in perfumes and for flavor.

BRADYKININ—also called kallidin—is a medical vasodilator. It is an experimental *teratogen*.

BROMOBENZENE—also called phenyl bromide—is an additive for motor oil. It is a skin *irritant*.

BROMOCIL is a *pesticide* (herbicide). It is an experimental *teratogen*. It is toxic by ingestion.

BROMOMETHANE—*see* methyl bromide.

BURNT LIME—*see* calcium oxide.

BUSULFAN is a *pesticide* (insecticide) and a medical *antineoplastic*. It is a *carcinogen*.

BUTACAINE is a medical and veterinary topical anesthetic. It is an *allergen*.

BUTALBITAL is a medical sedative. It is a depressant.

BUTALLYLONAL is a medical sedative. It is a central nervous system depressant and is poisonous.

1-BUTANAMINE—*see* n-butylamine.

n-BUTANOL—*see* n-butyl alcohol.

2-BUTANOL—*see* sec-butyl alcohol.

BUTETHAL is a medical sedative. It is a depressant.

BUTOXYETHANOL—*see* butyl cellosolve.

sec-BUTYL ACETATE is a lacquer solvent and solvent. It is an eye *irritant* and *central nervous system depressant.*

tert-BUTYL ACETATE is a lacquer solvent and gasoline additive. It is an eye and throat *irritant* and central nervous system depressant.

n-BUTYL ALCOHOL—also called n-butanol—is a lacquer solvent. It is a *central nervous system depressant* and an eye *irritant.*

sec-BUTYL ALCOHOL—also called 2-butanol—is used in polishes, paint remover, cleaning products, lacquer solvent, perfumes, and fruit essences. It is an eye *irritant*. Fatigue, headache and dizziness may result from inhalation.

tert-BUTYL ALCOHOL is used in lacquers, perfumes, fruit essences. It may have the same reactions as sec-butyl alcohol.

2-BUTYLAMINE—*see* sec-butylamine.

n-BUTYLAMINE—also called 1-butanamine—is used as an intermediate in the making of pharmaceuticals, *pesticide*s (insecticides), and synthetic tanning agents. It is a questionable *carcinogen* in laboratory tests. It is a potent eye, skin and mucous membrane *irritant*. It is a poison by ingestion.

sec-BUTYLAMINE—also 2-butylamine; Frucote; Tutane—is a *pesticide* (fungicide). It is a skin and mucous membrane *irritant*; toxic by ingestion.

BUTYLATED HYDROXYANISOLE—*see* BHA.

BUTYLATED HYDROXYTOLUENE—*see* BHT.

BUTYL CELLOSOLVE—also called 2-butoxyethanol—is a *pesticide*, is used as a solvent for resins, grease, oil, dry cleaning, and varnishes. There have been experimental *teratogenic* and reproductive effects in laboratory tests. It is an eye and skin *irritant*. It is toxic by inhalation, ingestion and skin contact.

BUTYLPHEN—*see* 4-tert-butylphenol.

4-tert-BUTYLPHENOL—also called butylphen—is used as a soap *antioxidant* and as an additive in motor oil.

BUTYROLACTONE—also called 3-hydroxybutyric acid lactone—is a product of paint removers. It is a skin *irritant*, toxic by ingestion, and is a suspected *carcinogen* in laboratory tests.

C

CACODYLIC ACID—also called dimethylarsenic acid—is a *pesticide*. It is a possible *carcinogen* and an experimental *teratogen*. It is an eye and skin *irritant*.

CADMINATE—*see* cadmium succinate.

CADMIUM is a powder used in mercury amalgam in dentistry. It is also used in veterinarian medicine. Its salts are used as anthelmintics in swine and poultry.

In humans it is a confirmed *carcinogen.* It is poisonous by inhalation.

CADMIUM SUCCINATE—also cadminate—is a *pesticide* (fungicide). It is a confirmed *carcinogen* and is poisonous by ingestion.

CADMIUM SULFIDE—also aurora yellow; C.I. Pigment Orange 20; C.I. Pigment Yellow 37—is a color for soaps, glass, textiles, and paper. It is a dermatological in medicine. It is a confirmed human *carcinogen.*

CAFFEINE is used as a human central nervous system stimulant. In veterinarian medicine it is used as a diuretic and cardiac and respiratory stimulant. It occurs naturally in coffee, tea and cola nuts. It is a *teratogen* and causes abnormalities of the craniofacial and musculoskeletal system.

CALCIUM BISULFATE—also calcium hydrogen sulfate—is a *pesticide* (germicide). It is a preservative, disinfectant, and fabric bleacher. It is a skin *irritant* and is poisonous by ingestion.

CALCIUM CARBONATE—also called chalk; dolomite; limestone; carbonic acid—is a *pesticide.* It is used in food, cosmetics, pharmaceuticals, and antibiotics. It is a medical antiacid and calcium supplement. It is used in cement. It is an eye and skin *irritant*.

CALCIUM CHLORIDE is a *pesticide*, a medical diuretic, and an antiallergenic. It is used in fire extinguishers and is a *wood preservative*. It is a questionable *carcinogen* in laboratory tests. It is toxic by ingestion.

CALCIUM CYANAMIDE—also called calcium carbimide—is a *pesticide* (defoliant, herbicide). It is a fertilizer and an *anthelmintic* in veterinarian medicine. It is a questionable *carcinogen* in laboratory tests. It is poisonous.

CALCIUM CYANIDE—also called cyanogas—is a *pesticide* (fumigant, rodenticide). It is a deadly poison.

CALCIUM FORMATE—also called formic acid—is used as a food and silage preserver. It is also used in lubricants. It is toxic by ingestion.

CALCIUM HYDROGEN SULFATE—*see* calcium bisulfate.

CALCIUM HYDROXIDE—also called slaked lime—is a *pesticide*. It is used as mortar, white wash, water softener, and as a *food additive*. It is a skin *irritant* and causes burns to the eyes.

CALCIUM OXIDE—also called burnt lime; quicklime—is a *pesticide* (fungicide, insecticide). It is used in bricks, plaster, mortar, and stucco. It is a clarifier for sugar cane and beet sugar juices, and is used to treat water and sewage. It is a strong *caustic* and may cause burns. It is an *irritant* to tissue. It is also a common air pollutant contaminant, and may ignite when mixed with ethanol.

CALCIUM PANTOTHENATE—also called pantothenic acid, calcium salt—is Vitamin B-5. It is a dietary supplement in veterinarian medicine and is toxic by ingestion.

CALCIUM PERMANGANATE is used as an antiseptic, disinfectant, and deodorizer. It is a strong oxidant. It may explode in presences with acetic acid or acetic anhydride. It may ignite when in contact with cellulose.

CALCIUM PHOSPHATE:

CALCIUM PHOSPHATE, DIBASIC—also called dicalcium orthophosphate—is used as a calcium supplement in foods. It is used in dental products. It is a fertilizer. It is an eye and skin *irritant*.

CALCIUM PHOSPHATE, TRIBASIC—also called tricalcium phosphate—is a *pesticide*. It is used in animal feeds and is a calcium supplement; it is also used in dental products. It is an eye and skin *irritant*.

CALCIUM PHOSPHIDE—also called tricalcium diphosphide—is a *pesticide* (rodenticide). It reacts with water to give off phosphine, a highly poisonous gas, which may ignite in air.

CAMOMILE—*see* chamomile.

CAMPHENE makes camphor, which is toxic by ingestion.

CAMPHOR is a *pesticide* (moth repellent). It is used in varnishes, lacquers, pharmaceuticals, and cosmetics. It is an eye and nose *irritant*.

CANDICIDIN—also Levorin—is used as a topical antifungal medication. It is toxic by ingestion and is an experimental *teratogen* in a laboratory test.

CANTHARIDES—also known as Spanish fly; blistering fly—is used as a medical *vesicant*. In veterinarian medicine it is used as a *vesicant* and *counterirritant*. It is toxic by ingestion. It can be *absorbed* through the skin and mucous membrane. It can cause death. Cantharides extracted from Spanish fly are *pesticides*.

CAPTAFOL—also called Difolatan; Folcid—is a *pesticide* (fungicide). It is a questionable *carcinogen* in a laboratory test, is an experimental *teratogen.* It is an *oncogen* and is toxic.

CAPTAN—also called Vancide—is a *pesticide* (fungicide). It is a bacteriostat in soap. It is toxic and may cause vomiting and/or diarrhea. It is a questionable *carcinogen* in a laboratory test.

CARAMEL is used in foods and mouthwash.

CARAWAY oil is found in the fruits of *Carum carvi L.* This oil contains carvone. Caraway oil is used as a spice. It is toxic by ingestion and skin contact, and is an *irritant* to skin.

CARBAMIDE—*see* urea.

CARBAMIDE PEROXIDE—also called urea peroxide—is used in s, pharmaceuticals, and tooth whiteners, and as a disinfectant. It is a strong oxidizing agent. It is an *irritant*.

CARBARYL—also called Sevin—is a *pesticide* (insecticide). It is toxic to bees and is a suspected human *carcinogen.*

CARBAZOLE—also called diphenylenimine—is a *pesticide* and a dye intermediate. It is a questionable *carcinogen* and is toxic by ingestion.

CARBENDAZIM—also called methyl 2-benzimidazolecarbamate—is a *pesticide.* It is a suspected *carcinogen.*

CARBOFURAN—also called Furdan—is a *pesticide* (insecticide, nematocide) and also kills birds. It is poisonous by inhalation, ingestion, and skin absorption. Laboratory tests indicate it is an experimental *teratogen.*

CARBON DIOXIDE is emitted from cars, power plants, and pressurized containers. It is an *asphyxiant* and central nervous system depressant. It has global warming effects and green house effects.

CARBON DISULFIDE is an air pollutant. It is a poison which is toxic by skin absorption. It may effect coronary heart disease. It effects menstrual flow and causes pain. It may cause miscarriages and it is a *fetotoxin.*

CARBONIC ACID—*see* calcium carbonate.

CARBON MONOXIDE is emitted from auto exhaust, gas (ovens, ranges, water heaters, dryers), wood stoves, coal burning, and fireplaces. It causes headaches, drowsiness, dizziness, and death.

CARBON TETRACHLORIDE—also tetrachloromethane—is a *pesticide* (grain fumigant), a solvent for fats, oils, varnishes, lacquers, degreaser, and cleaning. It is a central nervous system depressant. It can cause damage to the liver and kidney, is *carcinogen* and is a hepatotoxin. It breaks down to the poisonous gas phosgene.

CARBOXIN—also called Vitavax; Carboxin; DCMO—is a *pesticide* (fungicide). It is poisonous by ingestion.

CARRAGEEN—also called carrageenin; carra-geen—is a seaweed extract. It *absorbs* water, complexes with protein in milk, is an emulsifier; gelling agent; and a stabilizer in food products. It is used in cosmetics, pharmaceuticals, as an additive to hamburgers, and as a soothing ointment. If it complexes with milk protein it could effect mother's milk.

CARROTS and the parsley family (anise, caraway, dill) come from

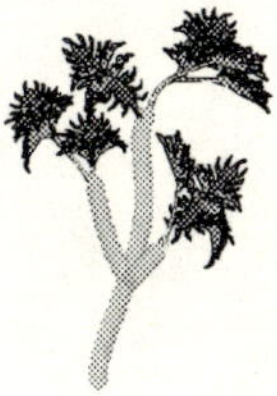

the umbelliferae family. Umbelliferae plants contain furanocoumarins which are *carcinogens*.

CARROT SEED OIL, which comes from the plant's seeds *Daucus carota L.*, is a skin *irritant*.

CARVAROL is naturally occurring in some plants. It is poisonous by ingestion and skin contact and is a skin *irritant*.

CARVONE is naturally occurring in some plants. It is used in liqueurs, flavor, *perfumes*, and soaps. It is toxic by ingestion.

CARYOPHYLLENE—also called beta-caryophyllene—is found in oil of clove, cinnamon leaves and other essential oils. It is used in *perfumes*. It is a skin *irritant*.

CASSIA OIL—*see* cinnamaldehyde.

CASTOR OIL—also called tangantangan oil—is from the seeds of *Ricinus communis L.* It is a medical cathartic, a veterinarian purgative, *emollient*, and a *pesticide*. It is toxic by ingestion, an allergen, and an eye *irritant*.

CAUSTIC POTASH—*see* potassium hydroxide.

CEDAR LEAF OIL—also called oil of cedar leaf—is obtained from leaves of *Thuja occidentalis L.* and is chemically identified as d-alpha-thujone. It is used medically as a *counterirritant* even though it is a skin *irritant*. It is toxic by ingestion and may cause death.

CEFOXITIN is used as medical *antibacterial*. If it gets into the body it can reduce white blood cell count.

CELLOPHANE is transparent and flexible. It is a questionable *carcinogen* in a laboratory test. When burned it can emit irritating fumes and smoke.

CELLOSOLVE—*see* 2-ethoxyethanol.

CELLULOID is used in the manufacture of toys, toilet products, surgical bandages, and in dentistry. It is toxic and *flammable*.

CETYL ALCOHOL—*pesticide*; shampoos; toxic by ingestion; eye and skin *irritant*.

CHALK—*see* calcium carbonate.

CHAMOMILE—also called camomile; ground apple—is from the dried flower heads of *Anthemis nobilis L.*, an aromatic bitter. It is an allergen.

CHENOPODIUM OIL—also called oil of American wormseed—contains ascaridol, cymene, camphor and saponins. It is poisonous by ingestion and skin contact, and it is a skin *irritant*.

CHERRY LAUREL OIL is a *volatile* oil from the leaves of *Prunus laurocerasus L.* It is very poisonous because of the release of hydrogen cyanide, especially when heated.

CHESTNUT TANNIN comes from chestnuts. It is a poison and a questionable *carcinogen* in a laboratory test.

CHLORAMBUCIL—also called chloroambucil—is a medical and veterinary *antineoplastic* and a *pesticide* (insect sterilant). It is a known human *carcinogen* causing leukemia and a poison. It affects the respiratory system and has had reproductive effects in the laboratory.

CHLORAMPHENICOL—is a medical *antibacterial* and antirickettsial; a veterinary antimicrobial; is a suspected human *carcinogen*; a poison.

CHLORAMPHENICOL PALMITATE is a medical *antibacterial* and antirickettsial, and a veterinary antimicrobial. It is a *teratogen* in a laboratory test.

CHLORDANE is a *pesticide* (insecticide, herbicide, termite control for homes). It is a *carcinogen.*

CHLORDIAZEPOXIDE HYDROCHLORIDE is a tranquilizer. It is a central nervous system depressant.

CHLORDIMEFORM—also called chlorophena-mine; Galecron—is a *pesticide* (acaricide, insecticide). It is poisonous by ingestion and skin contact. It is an eye and skin *irritant* and a questionable *carcinogen* in a laboratory test.

CHLORINATED LIME—also called bleaching powder—is a *pesticide*, a bleach, and is used to disinfect drinking water and sewage. It is a skin *irritant* and is toxic by inhalation.

CHLORINE is a *pesticide*, is used to purify water, and is a disinfectant. It is poisonous, is toxic by inhalation, is an eye and mucous membrane *irritant.* It reacts with other chemicals to form other *hazardous* chemicals.

CHLORNAPHAZINE is a medical *antineoplastic*. It is a known *carcinogen.*

CHLORPROMAZINE is an antipsychotic drug. It is toxic by ingestion.

CHLOROACETIC ACID is a *pesticide* (herbicide). It is an eye, skin and mucous membrane *irritant*. It is poisonous, and is a questionable *carcinogen* in a laboratory test.

alpha-CHLOROACETOPHENONE—also called chemical mace—is used as a riot control agent. It is an eye, throat and skin *irritant*. It is a poison by inhalation, and is a questionable *carcinogen* in a laboratory test.

CHLOROAMBUCIL—*see* chlorambucil.

CHLOROBENZENE is a paint solvent and is used in dry-cleaning. It is a skin *irritant* and central nervous system depressant.

CHLOROBENZILATE—also called Acaraben—is a *pesticide* (acaricide). It is a *carcinogen*, can cause testicular damage, and is toxic by ingestion.

ortho-CHLOROBENZYLIDENE MALONONITRILE is used in riot control. It is an eye, throat and skin *irritant*. It is poisonous by ingestion and toxic by inhalation. When heated, it can breakdown to poisonous cyanide gas.

4-CHLORO-META-CRESOL is used in medicine and veterinary medicine as a topical antiseptic. It is a poison and an allergen. When heated it can break down to toxic phosgene gas.

CHLOROETHANE—*see* ethyl chloride.

CHLOROFORM is a *pesticide*. It is used in drugs, cosmetics, food packaging, cough medication, and toothpaste, however, all these uses may not be availaboratoryle today. It is a *carcinogen*, central nervous system depressant, and hepatotoxin. It breaks down to phosgene, a poisonous gas.

CHLOROMETHANE—*see* methyl chloride.

3-CHLORO-2-METHYLPROPENE is a *pesticide* (insecticide, fumigant). It is an eye and respiratory *irritant*, toxic by inhalation, and a suspected *carcinogen* in laboratory test.

CHLOROPYRIFOS—also called Dursban—is a *pesticide* (insecticide, termite treatment). It is toxic and has central nervous system effects.

CHLOROTRIANISENE is a therapeutic estrogen used in the treatment of cancer. It is a suspected human *carcinogen*.

CHLOROXYLENOL—also called 4-chloro-3,5-xylenol—is used as an antiseptic, germicide, and *pesticide* (mildewcide). Medically it is used as an *antibacterial*, topical and urinary antiseptic, and as a veterinary topical antiseptic. It is toxic by ingestion, and is an experimental *teratogen*.

CHLORTETRACYCLINE—also called Aureomycin; Biomycin—is used as a medical *antibacterial* and antiamebic. In veterinary medicine it is used as an antimicrobial. This drug is a poison. It is toxic by ingestion and experimentally has shown reproductive effects.

CHOCOLATE contains theobromine,which is a poison. experimental *teratogen*. Chocolate is a diuretic, muscle relaxant, and bronchodilator.

CHOLEIC ACID—*see* deoxycholic acid.

CHOLESTEROL—also called cholesterin—is found in body tissue (fat, brain). It is used as an emulsifier in pharmaceutical and dermal preparation. It may contribute to heart problems, is an experimental *teratogen*, and a questionable *carcinogen* in a laboratory test.

CHROMIUM is used in autos. Some chromium compounds are *carcinogens*. It is *hazardous* waste.

CINNAMALDEHYDE—also called cinnamal—is found naturally in Ceylon and Chinese cinnamon oils. It is used as a flavor and in the perfume industries. It is toxic by ingestion and is *combustible*.

CINNAMAL—*see* cinnamaldehyde.

CINNAMYL ALCOHOL—also called styryl carbinol—is used in *perfumes* and as a deodorant. It is toxic by ingestion and is a skin *irritant*.

CINNAMON—*see* cinnamaldehyde.

CITRIC ACID is a *pesticide*. It is used in shampoos. It is toxic by ingestion, is an eye and skin *irritant*, and may cause allergies.

CITRONELLA OIL is a *pesticide* (insect repellant). It is used in soap perfume. It is not allowed on food crops.

CITRUS RED 2—also called Solvent Red 80; C.I. 12156—is used to color the skin of oranges. It is a suspected *carcinogen* in a laboratory test.

CLOVE LEAF OIL—also called oil of clove—is a *volatile* oil from *Eugenis caryopyllata Thunberg*. It contains eugenol, acetyleugenol, caryophyllene, furfural, vanillin, and methyl amyl ketone. It is used in toothpastes, confectionery, and medically as an anesthetic for toothaches. It is a skin *irritant* and is toxic by ingestion and skin contact.

COAL TAR PITCH is used in paints, roofing, and in paving roads. It contains acridine, anthracene, benzo (a) pyrene, fluorene, naphthalene, phenanthrene; some of these chemicals are *carcinogens*. *Coal tar pitch* is toxic by ingestion and causes cataracts.

COPAIBA OIL—also called oil of Copaiba—is used in *perfumes*. It is toxic by ingestion and can cause *dermatitis* and kidney damage.

COPPER is used in copper pipes, as a *pesticide* (insecticide); in paints; as an antifouling agent; in wiring; and to make alloys (brass,bronze). It is a questionable *carcinogen* in a laboratory test and is poisonous by ingestion.

COPPER ACETOARSENATE—also called paris green—is a *pesticide*. It is also used in *wood preservative* and antifouling paints.

COPPER ARSENATE is a *pesticide*.

COPPER ARSENITE is a *pesticide*. It is also used in pigment and in *wood preservative*.

COPPER SULFATE—also called blue stone—is a *pesticide* and a germicide. It is a soil additive and a *food additive*. It is toxic by ingestion, is a skin *irritant*, and is a questionable *carcinogen* in a laboratory test.

CORIANDER—also coriander oil—is from the fruit *Coriandrum sativiun L.* It resembles parsley, to which it is related. The leafs are used in salads and the seedlike fruit is ground for a flavoring. In addition, it is used as a carminative, an aromatic, and spice; and in gin. It is toxic by ingestion and is a skin *irritant.* It consists of malic acid, tannin, and mucilage.

CORN OIL—also called maize oil—is a product of corn. It is a cooking oil and is used in margarine and in enamel paints. It is also a pharmaceutical aid. It is a skin *irritant* and an experimental *teratogen* in laboratory test.

COTTONSEED OIL—is from the seed of *Gossypium herbarium L.* It is used in the making of oleomargarine, soaps, cosmetics, as a salad oil, and for cooking. It is a questionable *carcinogen* and a *teratogen* in a laboratory test.

COUMARIN—*see* Tonka bean.

CRESOL—also called cresylic acid and isomers (ortho, meta, para)—is a *pesticide* (insecticide). It is used as a disinfectant and antiseptic. It can burn eyes and skin, effects the central nervous system, and causes damage to kidney and liver.

CRESYLIC ACID—*see* cresol.

CROTONALDEHYDE—also called 2-butenal—is a *pesticide* (insecticide). It is a solvent in the purification of mineral oil and a warning agent in fuel gases. It is a suspected *carcinogen*, poisonous by inhalation and ingestion, and is an eye, skin and mucous membrane *irritant*.

CYANOGAS—*see* calcium cyanide.

CYGON—*see* dimethoate.

CYCLOHEXANE is a *pesticide* and a wax and oil solvent. It is an eye and mucous membrane *irritant*.

CYCLOHEXANONE is a *pesticide* (insecticide) and solvent for lacquers, waxes, and shellac. It is an eye and throat *irritant*, and has experimentally shown reproductive effects.

CYCOLPHOSPHAMIDE is a *pesticide* (insect *chemosterilant*), *antineoplastic* to treat leukemia, and is a known human *carcinogen*. It also has experimentally shown reproductive effects.

CYMENE—also called cymol—is a solvent. It is toxic by ingestion.

d

2,4-D—also called 2,4-dichlorophenoxyacetic acid—is a *pesticide* (herbicide). It is a *carcinogen*.

DAUNOMYCIN—is an antibiotic and *antineoplastic* (used to retard cancer). It is a *carcinogen*.

DBCP—*see* dibromochloropropane.

DCMO—*see* carboxin.

DDVP—*see* dichlorovos.

DEATH'S HEAD—*see* belladonna.

DECABORANE is a fuel additive, mothproofing agent, and propellent. It is poisonous by inhalation, ingestion and skin contact. It may explode with heat and ignites with oxygen.

DECAHYDRONAPHTHALENE—also DEC; Dec-alin; naphthalene—is a solvent for oils, fats, and waxes. It is also used in cleaning fluids, shoe cream, and floor waxes. It is toxic by inhalation and skin contact, is an eye and skin *irritant*, and can damage the kidneys. It is a questionable *carcinogen* in a laboratory test.

1-DECANAL—also called capraldehyde; capric aldehyde—occurs naturally in lemongrass, citronella, orange, and others. It is used in *perfumes* and as a flavor. It is toxic by ingestion and skin contact and is a skin *irritant*.

DECANE—also called decyl hydride—is a solvent. It is a questionable *carcinogen* in a laboratory test.

1-DECANOL—*see* decyl alcohol.

DECYL ALCOHOL—also called 1-decanol—is derived from coconut oil fatty acids, and is used in perfumes and as a flavoring agent. It is toxic by ingestion and skin contact and is an eye and skin *irritant*. It is a questionable *carcinogen* in a laboratory test.

DEET—*see* N,N-diethyl-m-toluamide.

DELNAV—*see* dioxathion.

DEMETON is a *pesticide* (insecticide) and cholinesterase inhibitor. It is toxic by ingestion, inhalation and skin contact, and has been fatal to humans.

DEOXYCHOLIC ACID—also called choleic acid—is used in medicine and as an emulsifying agent in foods. It is toxic by ingestion. It is a questionable *carcinogen* in a laboratory test.

DERRIS ELLIPTICA—also called derris root; tube root—is a *pesticide* (insecticide). It is poisonous by ingestion.

DEXTRAN—also called starch gum—is from glucose and consists of many different forms. It is a *food additive* and is used in blood plasma and in lacquers. Dextran(s) is a questionable *carcinogen* in a laboratory test.

DEXTROAMPHETAMINE SULFATE—also called dexamphetamine—is a central nervous system depressant treatment for humans and animals and an anorexic. It is a human *teratogen* and a poison.

DIACETONE—also called diacetone alcohol—is a solvent for lacquers, *wood preservatives*, stains, antifreeze, fats, and tars. It is toxic by ingestion, skin contact, and inhalation, and is an eye and mucous membrane *irritant*.

DIACETONE ALCOHOL—*see* diacetone.

DIALIFOR—also called Torak; dialifos—is a *pesticide* (insecticide, acaricide). It is poisonous by ingestion and skin contact.

DIALLATE—also called Avadex—is a *pesticide* (herbicide). It is toxic by ingestion and skin contact. It is a questionable *carcinogen* in laboratory test.

DIALLYL MALEATE—also called malic acid, diallyl maleate—is used in *pesticide* formulations. It is an eye and skin *irritant*, is toxic by skin contact, and is poisonous by ingestion.

DIAMINOZIDE—*see* alar.

DIAZINON is a *pesticide* (insecticide). It is poisonous by ingestion and skin absorption, and is an eye and skin *irritant*. It is an experimental *teratogen*.

DIAZOAMINOBENZENE—also called 1,3-diphenyltriazene—is a *pesticide* (insecticide) and is used in dyes. It is toxic by ingestion and is a questionable *carcinogen*.

DIBENZYLINE HYDROCHLORIDE is a medicine for an adrenergic blocker. It is toxic by ingestion. It is a suspected *carcinogen* and an experimental *teratogen.*

DIBROMOCHLOROPROPANE (DBCP) is a *pesticide* (nematocide). It is a *carcinogen* and can cause sterility.

DIBUCAINE is a medical anesthetic. It is poisonous by ingestion.

DIBUTYL SEBACATE (DBS) is used in cosmetics, perfumes, flavoring, and as a sealant for food containers. It is toxic by ingestion, and has experimental reproductive effects.

DICHLOFENITHION—also called Nemacide—is a *pesticide* (nematocide, insecticide). It is poisonous by ingestion and toxic by skin contact.

DICHLONE—also called Phygon—is a *pesticide* (fungicide, algicide, insecticide) and disinfectant. It is toxic by ingestion, inhalation and skin contact and is an eye, skin and mucous membrane *irritant.* It is a questionable *carcinogen* and can be a central nervous system depressant.

DICHLOROACETIC ACID is used in pharmaceuticals. It is toxic by ingestion and skin contact and is corrosive to the eyes, skin and mucous membrane. It is a questionable *carcinogen* in a laboratory test.

DICHLOROBENZALKONIUM CHLORIDE is a *pesticide* (germicide, algicide) and is used in sanitizer and deodorant. It is an eye *irritant*, poisonous if ingested, and can cause liver and kidney damage.

1,2-DICHLOROBENZENE—also called ortho-dichlorobenzene—is a *pesticide* (fumigant, insecticide) and a solvent. It is an eye *irritant*, central nervous system depressant, and can cause liver and kidney damage.

1,4-DICHLOROBENZENE—also called para-dichlorobenzene—is a *pesticide* (moth repellant, insecticide) and disinfectant. It is an eye and upper respiratory *irritant*, and is a potential *carcinogen*.

ortho-DICHLOROBENZENE—*see* 1,2-dichlorobenzene.

para-DICHLOROBENZENE—*see* 1,4-dichlorobenzene.

3,4-DICHLOROBENZIDINE is an intermediate for dyes and pigments. It is a confirmed *carcinogen*.

DICHLORODIFLUOROMETHANE also Freon 12 is a refrigerant. It destroys the ozone layer and is a central nervous system depressant.

1,3-DICHLORO-5,5-DIMETHYLHYDANTOIN (DDH)—also called Hydan; Amchlor—is a *pesticide* and is used in laundry bleach as well as in the treatment of water. It is toxic by inhalation and ingestion and is a skin *irritant.*

3,5-DICHLORO-N-(1,1-DIMETHYL-2-PROPYNYL) BENZAMIDE—also called Kerb; pronamide; propyzamide—is a *pesticide.* It is toxic by ingestion and a questionable *carcinogen.*

2,2'-DICHLOROETHYL ETHER—also called dichloroether—is a *pesticide* (soil fumigant) and solvent. It is toxic by inhalation, ingestion and skin absorption and is an eye, skin and mucous membrane *irritant.* It is a questionable *carcinogen.*

DICHLOROISOPROPYL ETHER is a solvent for fats, waxes, paints and varnish. It is used in cleaning. It is toxic by ingestion and inhalation and is an *irritant.*

DICHLOROMETHANE—*see* methylene chloride.

DICHLOROPHENARSINE HYDROCHLORIDE is a drug used to treat syphilis. It is toxic by ingestion. It contains arsenic and many arsenic chemicals are *carcinogens.*

2,4-DICHLOROPHENOXYACETIC ACID—*see* 2,4-D.

DICHLOROPROP—also called 2-(2,4-dichlorophenoxy) propionic acid; 2,4-DP—is a *pesticide* (herbicide). It is toxic by ingestion and skin contact. It is a suspected *carcinogen* and an experimental *carcinogen.*

1,2-DICHLOROPROPANE—also called propylene dichloride—is a *pesticide* (fumigant), stain remover, and solvent. It is an eye *irritant* and central nervous system depressant.

DICHLORO-S-TRIAZINETRIONE—also called dichloroisocyanuric acid—is a *pesticide* and is used in dry bleach, dish washing, and detergent sanitizer. It is toxic by ingestion and may cause stomach bleeding. It is an eye and skin *irritant.* It may ignite on contact with organic materials.

DICHLOROVOS—also called DDVP; Vapona—is a *pesticide* (fumigant, insecticide). It effects the central nervous system, is toxic by inhalation, ingestion and skin absorption, and is an *oncogen.*

DICYCLOHEXYLAMINE is a *pesticide* (insect-icide), *antioxidant*, catalyst for paints and varnishes, and is used in detergents. It is toxic by ingestion and is an eye, skin and mucous membrane *irritant*. It is a questionable carcin-ogen.

DIETHANOLAMINE (DEA) is used in cosmetics, cleansers, and detergents. It is an eye *irritant* and an animal *carcinogen*.

DIETHANOLAMINE (DEA) CETYL PHOSPHATE is used in hand and body lotions.

DIETHYL BENZENE is a solvent. It is an eye and skin *irritant*, toxic by ingestion and is a fire risk.

DIETHYLCARBAMAZINE CITRATE—also called Caritrol; Ditrazin; Ditrazine; Loxuran—is an anthelmintic. It is poisonous by inhalation and is an experimental *teratogen*.

DIETHYLENE ETHER—*see* dioxane.

DIETHYLENETRIAMINE (DETA) is a component in fuel. It is poisonous by skin contact; inhalation and ingestion, and is an eye and skin *irritant*.

DIETHYL GLYCOL (DEG)—also called glycol ether—is a solvent, antifreeze, and is used in cosmetics. It is toxic by ingestion and is an eye and skin *irritant*. It is a questionable *carcinogen*.

DIETHYLHEXYL ADIPATE (DEHA) is a plasticizer, solvent, and *emollient* in cosmetics. It is an animal *carcinogen*.

DI(2-ETHYLHEXYL) PHTHALATE (DEHP) is a plasticizer and a solvent. It may cause liver and testicular damage. It has *teratogenic* and *carcinogenic* effects in animals.

DI(2-ETHYLHEXYL) SODIUM SULFOSUCCINATE—also called dioctyl sodium sulfosuccinate—is a *food additive*, wetting agent, dispersant, and emulsifier. It is toxic by ingestion and is an eye and skin *irritant*.

DIETHYL KETONE—also called metacetone; propione—is toxic by ingestion.

7-DIETHYL-5-METHYL-S-TRIAZOLO (1,5-A) PYRIMIDINE—also called Rocornal; Trapymin—is a coronary vasodilator drug. It is poisonous by ingestion and is an experimental *teratogen*.

DIETHYL PHTHALATE (DEP)—also called Anazol—is a *pesticide* (insect spray, mosquito repellent), and is used in the manufacture of varnish, plasticizer, and as a solvent in *perfumes*. It is toxic by ingestion and inhalation and is an eye and mucous membrane *irritant*. It is an experimental *teratogen*.

DIETHYLSTILBESTROL is used in cattle growth hormones and to prevent spontaneous abortions. It is a confirmed *carcinogen* and a human *teratogen*.

N,N-DIETHYL-M-TOLUAMIDE—also called DEET; Off; Repel—is a *pesticide* (insecticide repellant) and a solvent. It is toxic by ingestion and inhalation, an eye and mucous membrane *irritant*, effects the central nervous system. It has shown experimental reproductive effects.

DIFOLATAN—*see* captafol.

DIHYDROCHOLESTEROL is a sterol found in feces. It is used in pharmaceutical prep. It is a questionable *carcinogen*.

l-DIHYDROXYPHENYLALANINE—also called l-dopa—Parkinson, is a disease drug. It is a questionable *carcinogen* and an experimental *teratogen*.

DIISOBUTYL KETONE—also called isovalerone—is a solvent. It is an eye and mucous membrane *irritant*.

DILL SEED OIL is naturally occurring. It is a skin *irritant*.

DIMETHISTERONE and ethinyl estradiol—also Ovin—is a steroid. It is a suspected *carcinogen*.

DIMETHICONE is a silicone oil and is used in shampoos.

DIMETHOATE—also called Cygon—is a *pesticide*. It is toxic by ingestion and skin contact. It is a questionable *carcinogen* and an experimental *teratogen*.

DIMETHOXYETHYL PHTHALATE (DMEP) is a plasticizer; and a solvent. It is a *teratogen* and has caused reproductive and fetotoxic effects in animals.

N,N-DIMETHYL ACETAMIDE—also DMA; DMAC—is a solvent and a paint remover. It is toxic by skin contact, inhalation and ingestion, and is an experimental *teratogen*.

DIMETHYLAMINE (DMA) is a *pesticide* (insecticide, fungicide), stabilizer in gasoline, propellant, and is used in soaps and surfactants. It is a fire risk and an eye, skin, mucous membrane and respiratory *irritant*.

DIMETHYLAMINOAZOBENZENE (DAB) is used in dyes, polishes, and waxes. It is toxic by inhalation and skin absorption, and is a confirmed/potent *carcinogen*. It is an experimental *teratogen*. It is prohibited from foods and beverages.

p-DIMETHYLAMINOBENZENEDIAZOSODIUM SULFONATE also Dexon—a *pesticide* (fumigant); questionable *carcinogen*; experimental *teratogen*; poisonous by ingestion.

2-DIMETHYLAMINOETHANOL (DMAE) also deanol—paint emulsifier; toxic by ingestion, inhalation and skin contact; eye and skin *irritant*; central nervous system stimulant.

N,N-DIMETHYLANILINE is a solvent. It is toxic by ingestion, inhalation and skin absorption, and it is a skin *irritant*.

DIMETHYLARSENIC ACID—*see* cacodylic acid.

1,1-DIMETHYLHYDRAZINE—also called dima-zine—is a *pesticide* (plant growth regulator) and is used in jet and rocket fuels. It is toxic by inhalation and by skin absorption and skin contact and is a confirmed *carcinogen* skin contact.

N,N-DIMETHYLFORMIDE (DMF) is a *pesticide* and a solvent. It is toxic by ingestion, skin absorption and inhalation, and is an eye and skin *irritant*. It is an experimental *teratogen*.

1,2-DIMETHYLHYDRAZONE is used in rockets. It is a confirmed *carcinogen* and is toxic by inhalation. It is an experimental *teratogen*.

1,2-DIMETHYLHYDRIZINE DIHYDROCHLORIDE is a rocket fuel. It is poisonous by ingestion and is a suspected *carcinogen*.

DIMETHYLOCTANOL—also called 3,7-dimethyl-1-octanol—*perfumes*; flavor; skin *irritant*; toxic by skin contact.

DIMETHYLOLUREA (DMU) is fire resistant, a plywood hardener, and is used in permanent-press fabrics. It is a skin *irritant*.

DIMETHYL PHTHALATE is a *pesticide* (insect repellent to be applied on skin), solvent, and plasticizer. It is an eye and mucous membrane *irritant*.

DIMETHYL SILICONE—also called dimethylpolysiloxane—is used in foods and as a defoaming agent. It is toxic.

DIMETHYL SULFOXIDE (DMSO) is a *pesticide*, medical and veterinary anti-inflammatory, solvent, and paint stripper. It is an eye and skin *irritant*, questionable *carcinogen*, experimental *teratogen*.

DIMILIN—also called Diflubenzuron—is a *pesticide*. It is toxic to aquatic invertebrates, may get into animal milk, and may cause *methemoglobinemia* and lung damage.

2,4-DINITROANILINE (DNA) is a pigment in printer ink toners. It is an eye *irritant* and is toxic by ingestion and inhalation. It is an experimental *teratogen*.

2,4-DINITROANISOLE is a *pesticide* (ovicide) and is used for body lice. It is toxic by ingestion and inhalation.

DINITRONAPHTHALENE and its isomers are used in dyes. It is a suspected *carcinogen*.

DINITROPHENOL and its isomers are used in dyes and as a lumber preservative. It is poisonous by ingestion and inhalation can be fatal. It is a skin *irritant* and an allergin.

DINITROTOLUENE and its isomers are confirmed *carcinogens*. It is toxic, an *irritant*, an allergin, and an experimental *teratogen*.

DINOSEB—*see* DNBP.

DIOXANE—also called diethylene ether—is a *pesticide* (fumigant) and is used in paints, solvent, varnish remover, deodorants, lacquers, and cosmetics. It is a fire risk, toxic by inhalation, toxic to the liver and kidney, an eye and mucous membrane *irritant*, and a *carcinogen* in animals.

DIOXATHION—also called Delnav—is a *pesticide*. It is poisonous by ingestion, inhalation, and skin contact.

DIPHACINONE—also called diphenadione—is a medical anticoagulant and a *pesticide* (rodenticide). It is toxic by ingestion and inhibits blood clotting.

DIPHENYLACETONITRILE—also called diphenatrile—is a *pesticide* (herbicide). It is toxic and is a questionable *carcinogen*.

DIPHENYLAMINE is a *pesticide* (preservative for stored apples) and is used in dyes. It is toxic by ingestion and is an experimental *teratogen*.

DIPHENYL CARBONATE is a solvent. It is a questionable *carcinogen*.

DIPHENYLENIMINE—*see* carbazole.

DIPHENYL OXIDE—also called diphenyl ether—is used in soaps and *perfumes*. It is toxic by inhalation.

DIPROPYL KETONE is used in solvent, lacquers, and flavoring. It is toxic by ingestion, inhalation, and skin contact.

DIQUAT is a *pesticide* (herbicide, plant growth regulator). It is an eye and skin *irritant*, toxic by ingestion, and is an experimental *teratogen*.

DISODIUM-3,6-ENDOXOHEXAHYDROPHTHAL-ATE—*see* accelerate.

DISODIUM METHANEARSONATE (DSMA) is a *pesticide*.

DISODIUM PYROPHOSPHATE—also called sodium pyrophosphate—is used in baking powder, cheese, meat, and desserts. It is an eye, skin and mucous membrane *irritant* and is toxic by ingestion.

DISODIUM SELENATE is a *pesticide*. It is a questionable *carcinogen* and an experimental *teratogen*.

DISULFIRAM is a *pesticide* (fungicide). It is an animal *teratogen*, toxic by ingestion, and is a questionable *carcinogen*.

DISULFOTON is a *pesticide* (insecticide, acaricide). It is toxic by ingestion, inhalation, and skin contact.

DIURON is a *pesticide* (herbicide). It is toxic by ingestion and is a questionable *carcinogen*.

DMSO—*see* dimethyl sulfoxide.

DNBP—also called Dinoseb (2-sec-Butyl-4,6-dinitrophenol)—is a *pesticide* (herbicide, insecticide, miticide). It is poisonous by any route of exposure and is a questionable *carcinogen* in a laboratory test.

1-DODECENE—*perfumes*—flavor; *irritant*.

DODECYL ALCOHOL—also called lauryl alcohol—is a *pesticide* and is used in perfumes as well as a flavoring agent. It is toxic by ingestion, a skin *irritant*, and is a questionable *carcinogen*.

DODECYLBENZENE is used in detergents. It is toxic by ingestion.

DOLOMITE—*see* calcium carbonate.

DRAMAMINE is a medication for motion sickness. It is poisonous by ingestion, effects perception, and can cause hallucinations.

DRAZOXOLON is a *pesticide* (antifungal). It is toxic by ingestion.

DULCIN is an artificial sweetener. Food use is prohibited by FDA.

DURSBAN—*see* chloropyrifos.

e

EDTA—also called ethylenediaminetetra-acetic—is a *pesticide* and is used in hair preparations. It is an experimental *teratogen*.

ENAVID—also called Enovid—is a steroid for birth control. It is a human *teratogen* and a confirmed *carcinogen.*

ENDOXAN—also called cyclophosphamide—is an *antineoplastic* and a sheep defleecing agent. It is poisonous, a confirmed human *carcinogen*, a skin *irritant*, and an experimental *teratogen.*

ENFLURANE is used as an anaesthetic. It is a questionable *carcinogen.*

EPHEDRINE is a bronchodilator. It is toxic by ingestion and raises pulse and blood pressure.

EPICHLOROHYDRIN is a *pesticide*. It is toxic, a *carcinogen*, and an *irritant*.

EPN is a *pesticide*. It is highly toxic.

ERYTHRITE is used in glass and ceramics. It is toxic by ingestion and inhalation.

ERYTHROMYCIN is an antibiotic. It is an experimental *teratogen*.

ESTRADIOL is a female sex hormone produced naturally. It is an *estrogenic* hormone. It is a *carcinogen*.

ESTRADIOL DIPROPIONATE is used as a menopause treatment. It is a confirmed *carcinogen* and an experimental *teratogen*.

ESTRIOL is a hormone. It is a *carcinogen*.

ESTROGEN is a natural female sex hormone. It is an oral contraceptive. It is a *carcinogen* and may cause *thromboembolism*.

ESTRONE is a natural female sex hormone. It relieves symptoms of menopause. It is a *carcinogen*.

ETHANOL—*see* ethyl alcohol.

ETHANOLAMINE is a *pesticide*, nonionic detergent, and is used in paints, polishes, and agricultural sprays. It is a skin *irritant*.

ETHINYL ESTRADIOL is a hormone. It is a *carcinogen* and is toxic by ingestion.

ETHION is a *pesticide*. It is poisonous by ingestion and is a skin *irritant*.

2-ETHOXYETHANOL—also called ethylene glycol monoethyl ether; Cellosolve—is used in cleaners, varnish remover, and is an anti-icing additive for aviation fuels. It is an eye *irritant*, toxic by inhalation and skin absorption, has caused lung and kidney damage in animals, effected bone marrow and testis; embryo/fetotoxicity effects; congenital malformation.

ETHYL ACETYL ACETATE is used as a flavor. It is an eye and skin *irritant* and is toxic by ingestion and inhalation.

ETHYL ALCOHOL—also called ethanol; grain alcohol—is a *pesticide*. It is also used in solvent, gasoline, beverages, and mouthwash. It is a confirmed *carcinogen*, a depressant, and an experimental *teratogen*.

ETHYL AMYL KETONE (EAK) is used in perfumes. It is an eye, skin, and mucous membrane *irritant*.

ETHYL ANTHRANILATE is used in perfumes and as a flavor. It is a skin *irritant* and is toxic by ingestion.

ETHYL BENZENE is a solvent. It is toxic by ingestion, inhalation and skin contact; an eye, skin, and mucous membrane *irritant*; and an experimental *teratogen*.

ETHYL BENZOATE is a solvent and is used in perfumes, and as a flavor. It is toxic by ingestion and is an eye and skin *irritant*.

ETHYL BROMIDE—also called bromoethane—is a *pesticide*, anesthetic, solvent, and is a refrigerant. It is toxic by ingestion, inhalation, and skin contact; an eye and skin *irritant*; and causes damage to liver and kidney.

ETHYL BUTYRATE—also called ethyl butanoate—is used in *perfumes*. It is an eye, skin, and mucous membrane *irritant*; and is toxic by ingestion.

ETHYL CAPRATE—also called ethyl decanoate—is used as a flavor. It is a skin *irritant*.

ETHYL CAPROATE—also called ethyl hexanoate—is a fruit essence. It is a skin *irritant*.

ETHYL CHLORIDE—also called chloroethane—is a *pesticide* and is used as a solvent. It is an eye, skin, and mucous membrane *irritant*. It is toxic by inhalation.

ETHYLENE—also called ethene—is a *pesticide*, refrigerant, and an anesthetic. It is *flammable*, an air contaminant, *asphyxiant*, and an explosive risk.

ETHYLENEDIAMINE—is a *pesticide* (fungicide) and a solvent. It is toxic by inhalation and skin absorption.

ETHYLENE DIBROMIDE (EDB)—is a *pesticide* (fungicide) and a solvent. It is toxic by ingestion and skin contact. and is a *carcinogen*. It is an experimental *teratogen*.

ETHYLENE DICHLORIDE (EDC) is a *pesticide* (fungicide) and a solvent. It is a *carcinogen* and an eye and skin *irritant*.

ETHYLENE GLYCOL is a *pesticide*, antifreeze and coolant, and is used in cosmetics, solvent, and de-icer for airport runways. It is toxic by ingestion and inhalation, and an eye, skin, and mucous membrane *irritant*. It is an experimental *teratogen*.

ETHYLENE OXIDE is a *pesticide* (fungicide). It is toxic by inhalation, a *carcinogen*, an eye and skin *irritant*, and is an experimental *teratogen*.

ETHYLENE THIOUREA—*see* 2-imidazolidi-nethione.

ETHYL FORMATE is a *pesticide* (fungicide and larvicide), solvent, and is used as a flavor. It is toxic by ingestion, inhalation, and skin contact, a skin *irritant*, and is a questionable *carcinogen*.

2-ETHYLHEXALDEHYDE is used in *perfumes*. It ignites in air.

2-ETHYL-1,3-HEXANEDIOL is a *pesticide* (insect repellent) and is used in cosmetics.

ETHYL HEXYLENE GLYCOL is a *pesticide* (insect repellent) and is used in hair preps. It is toxic by ingestion and skin contact, and is an eye and skin *irritant*.

ETHYL SILICATE is used in paints, lacquers, mortar, and cement waterproofer. It is toxic and is an eye, skin, and mucous membrane *irritant*.

ETHYL VANILLIN is a flavor. It is toxic by ingestion and is a skin *irritant*.

ETU—*see* 2-imidazolidinethione.

EUCALYPTOL is a naturally occurring oil. It is a cough medication, mouthwash, and is used in *perfumes*. It is poisonous by ingestion and is a skin *irritant*.

EUGENOL is used in perfumes, medicine, and as a flavor. It is toxic by ingestion, a skin *irritant*, and a questionable *carcinogen* in laboratory test.

EUGENOL ACETATE—*see* acetyleugenol.

FELDSPAR—different aluminosilicates (either barium, calcium, potassium, sodium)—is used in enamel, ceramics, pottery, concretes, fertilizer, and in tar roofing materials as an enamel. Powder of silicates can be toxic.

FENNEL OIL—also called oil of fennel—is toxic by ingestion and is a skin *irritant*.

FENTHION is a *pesticide* (insecticide; acaricide). It is toxic by ingestion and is a skin *irritant*. It is a questionable *carcinogen* and an experimental *teratogen*.

FERBAM is a *pesticide* (fungicide). It is toxic by ingestion, a questionable *carcinogen*, and an experimental *teratogen*.

FERRIC ARSENATE—*see* iron (III) arsenate.

FERRIC CHLORIDE—also called iron (II) chloride—is a *pesticide*, sewage treatment, feed additive, disinfectant for water, and is used in pigment. It is an eye and skin *irritant*, and is toxic by ingestion.

FERRIC OXIDE—*see* iron oxide.

FERROCENE—also called dicyclopentadienyl-iron—is an antiknock agent for gasoline and a fuel additive to improve combustion and reduce smoke. When heated it gives off toxic products which may be toxic by ingestion and inhalation. It is an *irritant* and may injure the liver and kidney.

FERROUS ARSENATE—*see* iron (II) arsenate.

FERROUS GLUCONATE—also called iron (III) gluconate—is a feed and *food additive*. It is toxic by ingestion a questionable *carcinogen*.

FERROUS SULFATE—also called iron (III) sulfate—is a *pesticide* (herbicide), water and sewage treatment, fertilizer, feed additive, *wood preservative*, and flour enrichment. It is poisonous, a questionable *carcinogen*, and an experimental *teratogen*.

FIR NEEDLE OIL is a flavor and a perfume odorant. It is a skin *irritant* and toxic by ingestion.

FLUORANTHENE comes from coal tar. It is a questionable *carcinogen.*

FLUORINE is used in toothpaste and a drinking water. It is a poisonous gas and is an eye, skin, mucous membrane, and tissue *irritant.*

FLUOROSILIC ACID—also called silicofluoric acid—is used in water purification. It is a poison, an eye, skin, and mucous membrane *irritant*, and reacts with water to produce toxic fumes.

5-FLUOROURACIL is an *antineoplastic* agent. It is a questionable *carcinogen* and an experimental *teratogen.*

FOLCID—*see* captafol.

FOLIC ACID is used in nutrition (Vitamin), medicine, and as a *food additive.* It is an experimental *teratogen.*

FORMAL—also called methylal; dimethoxymethane—is a solvent and a fuel. It is an eye and mucous membrane *irritant* and a central nervous system depressant.

FORMALDEHYDE is a *pesticide*, preservative, embalming fluid, foam insulation, and is used in plywood and furniture. It is a confirmed *carcinogen*, a poison, and an *irritant*.

FORMAMIDE is a solvent. It is an experimental *teratogen*, poison, and an eye *irritant*.

FORMIC ACID—*see* calcium formate.

FREON 12—*see* dichlorodiflurormethane.

FRUCOTE—*see* sec-butylamine.

FUMARIC ACID is a flavoring agent and is used in printing inks. It is an *irritant*.

FURAZOLIDONE is a *carcinogen*.

FURDAN—*see* carbofuran.

FURFURAL is a solvent, a flavor, and is used to make *pesticides*. It is an eye, skin, and mucous membrane *irritant*.

FUSEL OIL is a solvent and is used in *perfumes*. It is toxic and may contain chemicals which are *carcinogenic*.

G

GALACTOSE is a monosaccharide in milk and is used in medicine. It is an experimental *teratogen.*

GALLIC ACID is used in pharmaceuticals, writing ink, and is a tanning agent. It has experimental reproductive effects.

GELSEMINE is a medical central nervous system stimulant. It is a deadly poison and a toxic plant *alkaloid.*

GENTAMICIN is an antibiotic. It is an experimental *teratogen.*

GERANIOL is used in *perfumes*. It is a skin *irritant.*

GERANIUM OIL is a skin *irritant.*

GIBBERELLIC ACID is a *pesticide*. It is a questionable *carcinogen.*

GLUCONIC ACID—also called glycogenic acid—is used in food products, paint stripper, and pharmaceuticals.

GLYCERIN—also called glycerol—is a *pesticide* and hydraulic fluid; and is used in cosmetics, and is used in soaps, hydraulic fluid. It is toxic by ingestion, can cause headaches, nausea, vomiting, and is an eye and skin *irritant*.

GLYCEROL—*see* glycerin.

GLYCOGENIC—*see* gluconic acid.

GLYCOL DISTEARATE is used in shampoo. It is a type of ethylene glycol and many glycols are experimental *teratogens*.

GLYCOL STEARATE—also called ethylene glycol monostearate; stearic acid—is used in body and hand lotions and cosmetics. It is a skin *irritant*.

GOLD is used in dentistry and jewelry. It is a questionable *carcinogen*.

GRAPEFRUIT OIL is a skin *irritant* and a questionable *carcinogen* in laboratory experiments.

GRISEOFULVIN is a feed additive and an antifungal. It is a suspected *carcinogen*.

GUAIACOL—also called methylcatechol—is used in flavors. It is a poison, an eye *irritant*, and an experimental *teratogen*.

HALOTHANE is a medical anesthetic. It is similar to fluorocarbons (freons) which damages the ozone layer. It is a poison, an eye *irritant*, and an experimental *teratogen*.

HCN—*see* hydrogen cyanide.

HEMPA—also called hexamethylphosphoric triamide—*pesticide*; solvent; *carcinogen*.

HEPTACHLOR is a *pesticide* (insecticide). It contains chlordane, a *carcinogen*. Heptachlor is an *oncogen* and a *carcinogen*.

HEXACHLOROBENZENE—also called perchlorobenzene—is a *pesticide* and *wood preservative*. It is a confirmed *carcinogen* and an experimental *teratogen.*

HEXACHLOROPHENE is a *pesticide*, a topical anti-ineffective, germicide, and is used in soaps. It is a *carcinogen* and dioxin(S) are a byproduct.

HEXAMETHYL PHOSPHORAMIDE is a *pesticide.* It is a confirmed *carcinogen.*

HEXAMETHYLPHOSPHORIC TRIAMIDE—*see* hemps.

HEXANE—*see* n-hexane.

n-HEXANE—also called hexane—is a solvent and is used in the manufacture of glues and bandages. It is a central nervous system depressant and an *irritant* to the upper respiratory system.

sec-HEXYL ACETATE—also called methyl isoamyl acetate—is used in lacquers and fragrances. It is an eye and upper respiratory *irritant.*

HEXYLRESORCINOL is a topical antiseptic. It is an eye *irritant* and a questionable *carcinogen.*

HYDRALAZINE HYDROCHLORIDE is an *antihypertensive agent.* It is a suspected *carcinogen.*

HYDRATED SILICA—also called silica—is a *pesticide* and is used in toothpastes. It is toxic by inhalation. Dust can cause silicosis.

HYDRAZINE SULFATE is a *pesticide*. It is a *carcinogen*.

HYDROGEN CYANIDE (HCN)—also called hydrocyanic acid; formonitrile—is a *pesticide*. It is also found naturally in apricot seeds, peach seeds, leaves, tropical lima beans, in over 2,000 plant species. It is a poisonous gas. It is toxic by ingestion, inhalation, and skin absorption, has caused dyspnea, paralysis, headaches, vertigo, nausea, vomiting.

HYDROGEN PEROXIDE is a disinfectant, antiseptic, and mouthwash. Large amounts may cause death. In the body oxygen is released and could explode the body. Inhalation may cause dizziness, vomiting, convulsions, shock, and unconsciousness.

HYDROGEN SULFIDE—also called hydrosulfuric acid—is used around oil wells. It is given off by decaying plants (compost). It is an eye and respiratory *irritant*. A high level may cause respiratory paralysis and be fatal.

HYDROQUINONE is used in paint, varnish, fuel, and oil. It is an *irritant*, an *allergen*, and a questionable *carcinogen*.

HYPOGLYCIN A—also called alpha-amino-2-methylenecyclopropanepropanoic acid—is found in some plants. Unripe fruit can cause poisoning. Poison symptoms can be vomiting, coma, low blood sugar, and death.

2-IMIDAZOLIDINETHIONE—also called ethylene thiourea; ETU—is a confirmed *carcinogen*, an *irritant*, and an experimental *teratogen*.

INDOLE is used in *perfumes*. It is a *carcinogen*.

3-INDOLEACETIC ACID is a *pesticide* (plant hormone). It is a questionable *carcinogen* and an experimental *teratogen*.

INDOMETHACIN is a medicine for arthritis. It is a human *teratogen*.

IRON (II) ARSENATE—also ferrous arsenate—is a *pesticide*. It is a human *carcinogen* and a deadly poison.

IRON (III) ARSENATE—also ferric arsenate—is a *pesticide*. It is a human *carcinogen* and a deadly poison.

IRON DUST causes lung cancer.

IRON OXIDE—also called ferric oxide—is a paint pigment. It is a poison and a questionable *carcinogen*.

IRON OXIDE FUME is used in welding fumes. It is a questionable *carcinogen*.

ISOAMYL ACETATE—also banana oil—is a solvent and a flavor in water and syrups. It is an eye and mucous membrane *irritant*, and causes narcosis.

ISOAMYL BENZOATE is used in *perfumes*. It is a skin *irritant*.

ISOBUTYL ALCOHOL—also called isopropyl-carbinol—is a *pesticide* and is used in lacquer, paint remover, and in cleaner. Malignant tumors have been found in rats.

ISODRIN is a *pesticide*. It is toxic, causes acne, liver damage, and is a skin *irritant*.

ISOPHORONE—also I soforan—is a *pesticide* and lacquer solvent. It is an eye and mucous membrane *irritant*, and there have been preputial gland tumors. Kidney and lung damage can occur in animals.

n-ISOPROPYLACRYLAMIDE is used in detergents, cosmetics, and paper. If metabolized to acrylamide then a suspected *carcinogen* will be formed.

ISOPROPYL ALCOHOL—also called isopropanol; rubbing alcohol—is a *pesticide*, a solvent, and is used in skin lotions, cosmetics, and pharmaceuticals. It is an eye and mucous membrane *irritant*, a central nervous system depressant, and a questionable *carcinogen*.

ISOSAFROLE is used in *perfumes* and flavor, and as a *pesticide* synergist. It is a isosafrole contains safrole, and a suspected *carcinogen*.

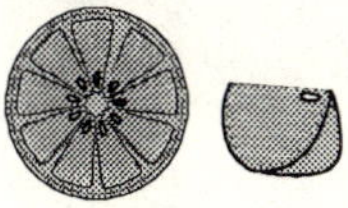

ISOVALERALDEHYDE occurs naturally in oranges and lemons. It is used in flavorings, perfumes, and may be metabolized to other aldehyde compounds which are *carcinogenic*.

J

JET FUEL HEF-2 is a poison.

JUNIPER BERRY OIL is a skin *irritant* and an *allergen.*

K

KALLIDIN—*see* bradykinin.

KARAYA GUM is used in food products. It is a mild *allergen.*

KEPONE is a *pesticide* (insecticide, fungicide). It is a *carcinogen* and *neurotoxin*, toxic by ingestion, inhalation, and skin absorption. It bioaccumulates in the food chain.

LACTOSE is used in infant foods, *perfumes*, and other processed foods. It has experimental reproductive effects, is a questionable *carcinogen* in experiments.

LANOLIN is a *pesticide* and a skin conditioner.

LATHYROGENS are amino acids in some plants and are toxic. They may be fatal. They cause muscular weakness and paralysis.

LAUGHING GAS—*see* nitrogen oxide.

LAUREL LEAF OIL is a skin *irritant*.

LAURIC ACID is a *pesticide*, *food additive*, and is used in cosmetics and soaps. It is a poison and a questionable *carcinogen* in experiments.

LAURYL ALCOHOL—*see* dodecyl alcohol.

LAVENDER OIL is used in *perfumes* and it gives off a pleasant odor. It contains other chemicals which may be a hazard, is a skin *irritant*, and is toxic by ingestion.

LEAD is used in storage batteries, paints, inks, ceramics, and bread wrappers. It causes neuromuscular dysfunctions, GI tract disturbances, and is a suspected *carcinogen*.

LEAD ARSENATE is a *pesticide* (insecticide, herbicide). It is used in waterproof paints and in antifouling paints. It is a *carcinogen* and is highly toxic.

LEAD NAPHTHENATE is a *pesticide* (insecticide, *wood preservative*), paint and varnish dryer, and additive for lube oil. It is a *carcinogen* and is toxic.

LEMONGRASS OIL is used in *perfumes* and as a flavor. It is a skin *irritant*.

LEMON OIL is a skin *irritant* and a questionable *carcinogen* in laboratory experiments.

LIMESTONE—*see* calcium carbonate.

LIMONENE—also called d-limonene—is a flavor, fragrance, and a solvent. It is toxic by ingestion and a questionable *carcinogen*.

LINALOOL—also linalol—is a flavor and is used in *perfumes*. It is toxic by ingestion and skin contact, and is a skin *irritant*.

LINDANE—also BHC; benzene hexachloride—is a *pesticide* (insecticide). It is an eye, nose, and throat *irritant*. It is excreted in milk, bioaccumulates in food chain, and is a *carcinogen*.

LINSEED OIL is used in paints and printing inks. It is an *allergen* and a skin *irritant*.

LITHIUM CARBONATE is used in ceramics, pharmaceuticals, in paint, varnish, and dyes. It is a human *carcinogen* and *teratogen*, and causes questionable leukemia and thyroid tumors.

MAGNESIUM ALUMINUM SILICATE is used in toothpaste.

MAGNESIUM CHLORIDE is used in fireproofing wood, ceramics, and is a *pesticide*. It is toxic by ingestion.

MAGNESIUM SILICATE is a *pesticide* (fungicide), preservative, *food additive*, and fertilizer. It is toxic by ingestion and is an experimental *teratogen.*

MAIZE OIL—*see* corn oil.

MALACHITE GREEN—also called benzal-dehyde—is a *pesticide* (fungicide) and an antiseptic. It is toxic by ingestion.

MALATHION is a *pesticide.* It is a poison and a questionable *carcinogen.*

MALEIC HYDRAZIDE (MH) is a *pesticide.* It is an *oncogen* and is toxic by ingestion.

MALIC ACID is a *pesticide.* It is toxic by ingestion and skin contact, and is an eye and skin *irritant.*

MALTOSE is a nutrient and a sweetener. It is an experimental *teratogen* and a questionable *carcinogen.*

MARJORAM OIL is used in soaps and toiletries. It is a skin *irritant.*

MECHLORETHAMINE—also called Nitromim—is used as a base for chemical warfare gas agent. It is a poison, a questionable *carcinogen* in a laboratory test, and an experimental *teratogen.*

MECHLORETHAMINE OXIDE HYDROCHLORIDE—also called oxyamine—is used as a medical *antineoplastic.* It is a poison and a suspected *carcinogen.*

MEK—*see* methyl ethyl ketone.

MENTHOL is a *pesticide*, is used in hand and body lotions, cough drops, cigarettes, chewing gum, and mouthwash. It is toxic by ingestion and an eye *irritant.*

MERCURY is a *pesticide*, is used in fillings, paints, and thermometers. It can cause death, psychological damage, and loss of teeth. It is an experimental *teratogen* and a questionable *carcinogen.*

MERCURIC ARSENITE is a *pesticide* and is used in waterproof paints and antifouling paints.

METALDEHYDE is a *pesticide* and a fuel. It is an *irritant*, poison, and has experimental reproductive effects.

METHANOL—*see* methyl alcohol.

METHAPYRILENE is an antihistamine. It is toxic by ingestion and is a *carcinogen.*

METHIONINE is used in feed additive, pharmaceuticals, and vegetable oil. It has experimental reproductive effects.

para-METHOXY ACETOPHENONE is used in *perfumes* and flavor. It is a skin *irritant* and can increase the pulse.

METHOXYCHLOR is a *pesticide.* It is a poison and a suspected *carcinogen.*

2-METHOXYETHANOL (2-ME)—also ethylene glycol monomethyl ether; methyl cellosolve—is a solvent and a anti-icing additive for jet fuel. It effects the central nervous system and causes reproductive effects (infertility, testicular atrophy, *teratogen*esis).

METHYL ACETATE—also acetic acid, methyl ester—is a solvent for lacquers. It is an eye and mucous membrane *irritant.*

METHYL ACRYLATE—also vitamin B1—is a poison and a questionable *carcinogen.*

METHYL ALCOHOL—also methanol—is a *pesticide*, antifreeze, and a fuel. It is a poison, an *irritant*, and an experimental *teratogen*.

METHYL AMYL ALCOHOL is used in *perfumes*, as a flavor, and a solvent. It is toxic by ingestion, inhalation, and skin contact, and is a skin *irritant*.

METHYL 2-BENZIMIDAZOLECARBAMATE.

METHYL BROMIDE—also bromomethane—is a *pesticide* (soil and stored food fumigant), refrigerant, and fire-extinguishing agent. It is a *neurotoxin*, can cause convulsions, and is a suspected *carcinogen*.

METHYLCATECHOL—*see* guaiacol.

METHYL CELLOSOLVE—also 2-methoxyethanol; (2-ME)—is a *pesticide*, solvent, is used in *perfumes*, and as a de-icer for jet fuel. It is a *teratogen*, causes testicle atrophy, is toxic by ingestion, inhalation, and skin contact, and is an eye and skin *irritant*.

METHYL CHLORIDE—also chloromethane—is a *pesticide* (fungicide), and an anti-knock fuel additive. It is a central nervous system depressant, *carcinogen* in laboratory animals, can cause liver and kidney damage, and is an experimental *teratogen*.

METHYLCHLOROFORM—*see* 1,1,1-trichloroeth-ane.

METHYLENE CHLORIDE—also called dichloromethane—is a paint remover, solvent, eye, skin, and respiratory tract, central nervous system depressant, and a *carcinogen* in laboratory animals.

METHYL ETHYL KETONE—also called MEK—is a solvent and a *pesticide*. It is toxic by inhalation and an *irritant*.

METHYL EUGENOL is a *pesticide* (insect attractant) and a flavor.

METHYL GUTHION—*see* azinphos methyl.

6-METHYL-5-HEPTEN-2-ONE is used in *perfumes* and as a flavor. It is a skin *irritant*.

METHYL HYDRAZINE is a solvent. It is a *carcinogen*.

METHYLPARABEN is a *pesticide* and is used in hair preparations.

METHYL PARATHION is a *pesticide*, a poison, a questionable *carcinogen*, and an experimental *teratogen*.

METHYL SALICYLATE is a *pesticide* and is used in beverages, foods, and mouthwash. It is a poison by ingestion, an eye and skin *irritant*, and an experimental *teratogen*.

MICA has a number of silicates. It is used in exterior paints, roofing compositions, wall-board joint cement, and cosmetics. Its dust is an *irritant* and when inhaled it may cause lung damage.

MINERAL OIL—also paraffin oil; petrolatum liquid—is a *pesticide*, lubricant, ink solvent, medical cathartic, and laxative. Some mineral oils in the past contained *carcinogens*. It is a human *teratogen* (causes testicular tumors in the fetus).

MINERAL PITCH—*see* asphalt.

MINERAL SPIRITS—*see* stoddard solvent.

MONOAMMONIUM METHANEARSONATE is a *pesticide*.

MONOSODIUM ACID METHANEARSONATE is a *pesticide*.

MONOSODIUM GLUTAMATE—also MSG—causes hallucinations and is an experimental *teratogen*.

MORPHOLINE—also diethyleneimide—is a *pesticide*, wax solvent, and an antiseptic. It is an eye, nose, and throat *irritant*, and may be metabolized to nitroso compounds which are *carcinogens*.

MSG—*see* monosodium glutamate.

MUCILAGE is a plant part.

MYRCENE is used as a flavor and in the manufacture of perfumes.

NAPHTHA is a solvent. It is a poison.

NAPHTHALENE is a *pesticide*, preservative, and a lubricant. It is a poison, an *irritant*, and an experimental reproductive effects.

NIACIN—also called nicotinic acid—is a *pesticide*, dietary supplement, and medicine. It is toxic, causes liver damage, and a questionable *carcinogen* in experiments.

NICOTINE—also 1-methyl-2-(3-pyridyl) pyrrolidine—is a *pesticide* (insecticide). It is a rapid-acting poison, effects the central nervous system, may cause nausea, vomiting, dizziness, cardiac arrhythmias, death. It is a *carcinogen* and a *teratogen* in laboratory animals.

NICOTINIC ACID—*see* niacin.

NITRILOTRIACETIC ACID (NTA)—also called triglycine—is used to make synthetic detergents. It is a *carcinogen* in laboratory animals.

NITROBENZENE—also called oil of mirbane—is a *pesticide*, and is used in metal polishes and shoe polishes. It is toxic by ingestion, inhalation, and skin absorption, and has experimental reproductive effects.

4-NITRODIPHENYL—also called para-nitrobi-phenyl (PBN)—is a *pesticide* (fungicide), plasticizer, and *wood preservative*. It is a *carcinogen* and causes bladder cancer.

NITROGEN OXIDE—also called laughing gas—is an experimental *teratogen* and is moderately toxic by inhalation.

1-NITRONAPHTHALENE is added to mineral oils and dyes. It is an *irritant* and a suspected *carcinogen.*

N-NITROSODIMETHYLAMINE—also called dimethylnitrosamine (DMN)—is a solvent, and is often found in fish meal. It is a *carcinogen.*

NUTMEG OIL—also called oil of nutmeg—is a skin *irritant* and has experimental reproductive effects.

NUTRASWEET—*see* aspartame.

NYLON is a questionable *carcinogen.*

OIL of AMERICAN WORMSEED—*see* chenopodium.

OIL of BASIL—*see* basil oil.

OIL of BAY—*see* bay oil.

OIL of CEDAR LEAF—*see* cedar leaf oil.

OIL of CLOVE—*see* clove leaf oil.

OIL of MIRBANE—*see* nitrobenzene.

OIL of NUTMEG—*see* nutmeg oil.

OREGANO—also called marjoram—is an herb. It is used a a food flavoring. Its oil is also used as a flavoring agent as well as in perfumes.

ORTHOARSENIC ACID—*see* arsenic acid.

OVIN—*see* dimethisterone.

OXALIC ACID is a *pesticide*, which is used in bleaches, rust remover, and metal polishes. It is an eye, skin, and mucous membrane *irritant*.

OZONE is a poison, an *irritant*, an experimental *teratogen*, and a questionable *carcinogen* in experiments.

PALMITIC ACID is used in soap and in food grade additive. It is a poison, an *irritant*, and a questionable *carcinogen* in experiments.

PANTOTHENIC ACID—*see* calcium pantothenate.

PAPAIN is used in cosmetics, pharmaceuticals, and as a meat tenderizer. It is an *allergen* and an experimental *teratogen*.

PARAFFIN WAX is a mixture of *hydrocarbons*. It is used in wax paper, candles, polishes, beverages, s, cosmetics, and surgery. Many paraffin waxes are *carcinogenic*.

PARAFORMALDEHYDE is a *pesticide* (fungicide, bactericide), disinfectant, and contraceptive cream. It is toxic by ingestion. Polymerized formaldehyde and formaldehyde are *carcinogens*.

PARSLEY—also oil of parsley—occurs naturally and is used with foods. It contains many chemicals, is toxic by ingestion, and is a skin *irritant*.

PCBs (polychlorinated biphenyls) are a *pesticide* and a fire retardant in electric generators. They are a *carcinogen* and they bioaccumulate in the food chain.

PCP—*see* pentachlorophenol.

PEG 8 is a polyethylene glycol series of carbowax chemicals. Some are toxic, *irritant*s, and questionable *carcinogens*.

PENICILLIN G. SODIUM SALT—*see* benzyl pencillin sodium.

PENTA—*see* pentachlorophenol.

PENTACHLOROPHENOL—also called penta; PCP—is a *pesticide* (defoliant, herbicide, *wood preservative*, termiticide). It contains impurities known as dioxin(s). It is a *carcinogen*, and is an eye and upper respiratory *irritant*.

PERCHLOROBENZENE—*see* hexachlorobenzene.

PERCHLOROETHYLENE—also called tetrachloroethylene—is a *pesticide*, solvent, and cleaning agent. It is a central nervous system depressant, can cause liver damage, and a *carcinogen* in a laboratory test. Do not confuse with trichloroethylene (TCE).

PERTHANE—also called diethyldiphenyl dichloroethane—is a *pesticide* (insecticide). It is poisonous by ingestion, a questionable *carcinogen*, and an experimental *teratogen*.

PEPPERMINT OIL contains menthol and other chemicals. It enhances flavors, is used in chewing gum, liqueurs, medicines, mouthwash, and . It is toxic by ingestion and an *allergen*.

PETROLEUM SPIRITS—*see* stoddard solvent.

PETROLATUM is used in hand and body lotions. Many are *carcinogens*.

alpha-PHELLANDRENE is a natural chemical found in eucalyptus oil and cinnamon oil. It is used as a flavor and in *perfumes*. It is toxic by ingestion and skin absorption, and is an *irritant*.

PHENOBARBITAL is a drug to treat epilepsy and a sedative. It is a suspected *carcinogen* and a human *teratogen.*

PHENOL—also called benzophenol; carbolic acid; hydroxy benzene; phenylic acid—is a *pesticide*, antiseptic, disinfectant, and sore throat medication. It is toxic by ingestion, inhalation, and skin absorption, an eye, skin, and mucous membrane *irritant*, and can cause liver and kidney damage. It may be a *fetotoxin.*

PHENYL BROMIDE—*see* bromobenzene.

para-PHENYLENEDIAMINE—also called para-diaminobenzene—is used in hair-dye products. It may cause bronchial asthma and other respiratory problems.

2-PHENYLETHANOL (PEA)—also called benzyl carbinol—is a *pesticide*, *food additive*, fragrance, preservative, and an antimicrobial agent. It is an eye *irritant* and a *teratogen* in laboratory animals.

PHENYLHYDRAZINE is used in the manufacture of pharmaceuticals. It is a *carcinogen.*

PHOSGENE is a poison, toxic gas, *irritant*, and a breakdown product of many chemicals (e.g., chloroform).

PHOSPHENE is a *pesticide*. It is a poison, toxic gas, and a breakdown product of many chemicals.

PHOSPHORIC ACID is a *pesticide* and fertilizer. It is used in soaps, detergents, foods, and beverages. It is a poison and a corrosive *irritant*.

1-PINENE is found in spices such as thyme oil, and is an ingredient in perfumes and flavors.

2-PINENE is found in spices such as thyme oil. It is a skin *irritant*.

PINE OIL is a *pesticide*, preservative, and a fragrance. It is an *allergen*, *irritant*, and is toxic.

POISON BLACK CHERRY—*see* belladonna.

PORTLAND CEMENT may contain chromium and some chromium compounds are *carcinogens*. It is an eye *irritant* and can cause *dermatitis*.

POTASSIUM ARSENATE is a *pesticide* and is used in fly paper.

POTASSIUM HYDROXIDE—also called *caustic* potash—is used in the manufacture of liquid and soft soaps. It is an eye, skin, and mucous membrane *irritant*, and can cause burns.

POTATOES, like apples, contain a chemical in the skin called solanine. Solanum *alkaloids* have resulted in human poisonings and death, caused headaches, diarrhea, fatigue, nausea, and vomiting. Solanum has been used as a *pesticide* (insecticide).

PROGESTERONE is a natural female sex hormone. It is an oral contraceptive. It is a confirmed *carcinogen.*

PROPENAMIDE—*see* acrylamide.

2-PROPENETRILE—*see* acrylonitrile.

beta-PROPIOLACTONE (BPL) is a *pesticide*, disinfectant, and sterilant. It is a skin *irritant* and a *carcinogen.*

PROPYL ACETATE is a solvent and is used in perfumes. It is an eye and mucous membrane *irritant.*

PROPYLENE GLYCOL is an antifreeze, *pesticide*, wax and oil solvent, de-icer for airport runways, moisturizing cream, cleansing cream, ingredient in sun-tan lotions, shampoos, deodorants, baby lotions, and wash towlettes. It is an eye and skin *irritant* and an experimental *teratogen*.

PROPYLENE IMINE is an intermediate in the manufacture of paper finishes, adhesives, and coatings. It is an animal *carcinogen* and an eye *irritant*.

PROPYLENE OXIDE is a *pesticide*, solvent, and is used in surfactant and detergents. It is a confirmed *carcinogen*, an *irritant*, and has experimental reproductive effects.

PYRETHRUM is a group of chrysanthemum flowers that yield pyrethrum extract. The extract is a *pesticide* (insecticide) and medication for toothaches and headaches. It can cause *dermatitis*, and if ingested can cause vomiting, *bradycardia*, and destruction of the liver.

Q

QUICKLIME—*see* calcium oxide.

RADON is a gas found in soil, underground water (well water), air, rocks. It is a *carcinogen* causing lung cancer.

REDSKIN—*see* allyl isothiocyanate.

RESORCINOL is used in the manufacture of adhesives, cosmetics, dyes, and *pesticide*. It is an eye and skin *irritant* and a questionable *carcinogen* in experiments.

RESERPINE is an *antihypertensive agent* and a veterinary tranquilizer. It is a *carcinogen*.

ROSEMARY OIL is naturally occurring and contains many other chemicals which may be *hazardous*. It is toxic by ingestion and is a skin *irritant*.

ROTENONE—also called tubatoxin—is a *pesticide*(insecticide) and a lotion for chiggers. It affects the central nervous system and is a *carcinogen*.

SACCHARIN is a sweetener commonly used as a sugar substitute. It is a *carcinogen.*

SAFROLE naturally occurs in some plants. It is a *pesticide* (mosquito repellent), medical antiseptic, and is used in perfumes and soaps. At one time it was used in soft drinks and foods. It is a *carcinogen* and is toxic by ingestion.

SAGE OIL is derived from a naturally occurring plant source. It is used in *perfumes* and as a flavor. It contains other chemicals which may be *hazardous*. It is a skin *irritant*; and in some applications toxic by ingestion.

SASSAFRAS is a skin *irritant* and a questionable *carcinogen* in experiments.

SAVORY OIL is a naturally occurring herb. It is poisonous by skin contact, toxic by ingestion, and is a skin *irritant.*

SELENIUM is naturally occurring. It is a *pesticide* and a vitamin. Some selenium compounds are *hazardous*. It is a questionable *carcinogen* and has experimental reproductive effects.

SELENIUM SULFIDE—also selenium disulfide—medicates shampoo. It is toxic by ingestion, eye, and skin *irritant*, and is a *carcinogen* to animals.

SEVIN—*see* carbaryl.

SILVER NITRATE is a *pesticide*, antiseptic, and a hair dye. It is a *caustic*, poison, an *irritant*, and an experimental *teratogen* in experiments.

SKATOLE—also called 3-methylindole—is naturally occurring in feces, *civet* cat, and coal oil. It is used in perfumes. Many indole compounds are *carcinogens*.

SOAP may be sodium tallowate, sodium cocoate and/or sodium palm kernelate (present in varying percentages). It is used as a *pesticide*.

SOAPSTONE is naturally occurring. It is used in paint pigment and soaps. Talc fibers (if present) may cause lung cancer.

SODA ALUM—*see* aluminum sodium sulfate.

SODIUM ACETATE is a *pesticide* and is used in soaps, medicine, and as a *food additive*. It is a poison and is an *irritant*.

SODIUM BENZOATE is a *pesticide*, food preservative, and is used in tobacco. It is a poison and is an experimental *teratogen*.

SODIUM CITRATE is used in shampoos, foods, and detergents.

SODIUM CHROMATE is a *pesticide* and paint pigment. It is a confirmed *carcinogen*.

SODIUM CYCLAMATE is an artificial sweetener. It is a *carcinogen*.

SODIUM DICHLOROISOCYANURATE is a *pesticide*, swimming pool disinfectant, is used in dry bleaches, and to treat water and sewage. It is toxic by ingestion and can be a fire risk. Do not use in well water.

SODIUM FLUORIDE is a *pesticide* (insecticide, fungicide, rodenticide), *wood preservative*, water purifier, disinfectant, and is used in toothpastes. It is toxic by ingestion and a tissue *irritant*.

SODIUM HYPOCHLORITE is a *pesticide*, swimming pool disinfectant, and water purifier. It is toxic and an *irritant*.

SODIUM LAURYL PHOSPHATE is a *pesticide* and is used in toothpastes.

SODIUM LAURYL SULFATE is a *pesticide*, *food additive*, and is used in toothpaste. It is a skin *irritant*, an *allergen*, and an experimental *teratogen*.

SODIUM PROPIONATE is a *pesticide* and food preservative. It is an *allergen*.

SODIUM PYROPHOSPHATE—*see* disodium pyrophosphate.

SODIUM SACCHARIN is a sweetener. It is a *carcinogen*.

SODIUM SILICOFLUORIDE—also called sodium fluorosilicate—is a *pesticide* (insecticide, rodenticide) and *wood preservative*.

SODIUM ZIRCONIUM GLYCOLATE is a deodorant, *astringent*, and germicide. Zirconium is prohibited from cosmetics—it is a *carcinogen*.

SODIUM ARSENATE is a *pesticide* and is used in dyeing and printing.

SODIUM ARSENITE is a *pesticide*, antiseptic, and is used in dyeing.

SODIUM ZIRCONIUM LACTATE is used in deodorants and antiperspirants. Zirconium is prohibited from cosmetics—it is a *carcinogen.*

SOLANINE—*see* potatoes.

SOLAVUM—*see* potatoes.

SORBIC ACID is a *pesticide* and food preservative. It has experimental reproductive effects and is a questionable *carcinogen* in experiments.

SORBITOL is used in toothpastes. It is toxic by ingestion, may cause *hypermotility* and diarrhea.

SPANISH FLY—*see* cantharides.

SPEARMINT OIL is naturally occurring and contains other chemicals which may be *hazardous*. It is toxic by ingestion, an *allergen*, and a skin *irritant.*

SPIRONOLACTONE is a diuretic. It is a *carcinogen* in laboratory animals.

STANNIC CHLORIDE—also called tin chloride—is used in soaps. It is toxic by inhalation, and is an eye, skin, and mucous membrane *irritant*.

STANNOUS FLUORIDE—also called tin fluoride—is used in toothpastes. It is toxic by ingestion and is an *irritant* to tissue.

STARCH GUM—*see* dextran.

STEARIC ACID is naturally occurring in some animals and plant fats. It is poisonous, a skin *irritant*, and a questionable *carcinogen* in a laboratory test.

STEARYL ALCOHOL—also called 1-octadec-anol—is used in shampoos. It is toxic by ingestion and is a questionable *carcinogen* in a laboratory test.

STODDARD SOLVENT—also called mineral spirits; petroleum spirits—is a mixture of aromatic *hydrocarbons*, paraffin, and naphthenics. It is a *pesticide*, is used in dry cleaning, as a degreaser, a spot and stain remover, and a paint thinner. It is an eye *irritant* and central nervous system depressant. Many paraffins are *carcinogens*.

STRONTIUM-90 is present in nuclear fallout. It has been found in milk (cows, etc.). It attacks bone marrow.

SULFUR DIOXIDE is a *pesticide*. It is an air pollutant from discharges and reacts with rain (water) to form acid rain (sulfuric acid). It can cause chronic bronchitis and respiratory problems, is an eye, skin, and mucous membrane *irritant*, and is *hazardous* to the environment.

SULFURIC ACID is a *pesticide*, fertilizer, and a laboratory *reagent*. It is a corrosive, an *irritant* to tissues and eyes, and can cause inflammation of the upper respiratory tract.

SULFUROUS ACID is a preservative for foods, nuts, and wines. It is toxic by ingestion and inhalation, and is an *irritant* to tissue.

SWEET ALMOND is the rape seed of *Prunus amygdalus Stokes var dulcis (D.C.) Baill*; etc. It is used in perfumes, confectioner, almond milk, and meal.

t

TALC is used in cosmetics, soap, paints, crayons, and plaster. It is an *irritant*, toxic by inhalation, a questionable *carcinogen* in experiments, and some talc may contain asbestos.

TANGANTANGAN OIL—*see* castor oil.

TANGERINE OIL is a skin *irritant*.

TANNIC ACID—also called tannin—is naturally occurring. It is a *pesticide*, and occurs in wines, brewing, foods, tea, and coffee. It is toxic by ingestion and inhalation. Tannins derived from nutgall may be *carcinogenic*.

TANNIN—*see* tannic acid.

TARRAGON OIL—also called estragole; methyl chavicol—is naturally occurring. It is used in perfumes and as a flavoring.

TARRAGON—*see* para-allylanisole.

TCE—*see* 1,1,1-trichloroethane.

TEA—*see* triethanolamine.

TEMIK—*see* aldicarb.

TERPENE(s) occurs naturally in lemon, caraway, orange, peppermint, and spearmint.

TERPINEOL(s) is naturally occurring in spices/herbs. It is a *pesticide*. It is toxic by ingestion and is a skin *irritant*.

TESTOSTERONE is a hormone and is used in medicine. It is a confirmed *carcinogen*.

TETRACHLOROETHYLENE—*see* perchloroethylene.

TETRACHLOROMETHANE—*see* carbon tetrachloride.

TETRAETHYL LEAD (TEL) is an antiknock for gasoline. It effects the nervous system, can cause personality changes, and acute and chronic poisoning.

TETRAETHYLTHIURAM SULFIDE is a *pesticide* (fungicide, insecticide) and is used in pharmaceutical ointments. It is toxic by ingestion.

TETRAHYDRONAPHTHALENE is a solvent for fats, grease, oils, and waxes. It is an eye and skin *irritant*.

TETRAMETHYL LEAD (TML) is used in gasoline and aviation fuels. It effects the nervous system.

TETRAPHENE—*see* 1,2-benzanthracene.

TETRAPOTASSIUM PYROPHOSPHATE (TKPP) is used in soaps and detergents.

TETRASODIUM PYROPHOSPHATE (TSPP) is a water softener. It is used in soaps, detergents, metal cleaner, and as a *food additive*. It is toxic by inhalation and poisonous by ingestion.

TETRASODIUM ETHYLENEDIAMINETETRAACETIC ACID (EDTA)—is used in soap.

THALLIUM OXIDE—also called thallus oxide—is used in synthetic gemstones and optical glass. It is toxic by ingestion.

THALIDOMIDE is a human *teratogen*.

THEOBROMINE is found in chocolate. It effects the central nervous system, is a poison, and is an experimental *teratogen*.

THYMOL is a *pesticide* and preservative. It is used in *perfumes* and mouthwash. It is poisonous by ingestion, an *allergen*, and has reproductive effects in a laboratory test.

THIOUREA—also called thiocarbamide—is a synthesis of dyes, drugs, and hair preparations. It is a skin *irritant* and *carcinogen*.

THUJONE is naturally occurring in sage oil. It is a spice, an herb, and a solvent. It is toxic by ingestion.

THYME OIL—also oil of Thyme; thym oil—is naturally occurring in spices and herbs. It contains many other chemicals which may be *hazardous*. It is toxic by ingestion, a skin *irritant*, and an *allergen*.

TIN CHLORIDE—*see* stannic chloride.

TINCTURE of IODINE is a *pesticide* and antiseptic for skin surfaces. It is toxic by ingestion.

TIN FLUORIDE—*see* stannous fluoride.

TITANIUM DIOXIDE is used in paints, ceramics, and toothpastes. It is a pulmonary and skin *irritant*, and a questionable *carcinogen.*

TNT—also called 2,4,6-Trinitrotoluene (TNT) is an explosive. It can cause liver damage, and could be a carcinoma of the urinary bladder.

TOCOPERYL ACETATE—also called Vitamin E acetate—is used in hand and body lotions. Vitamin E has shown reproductive effects in laboratory test.

TOLUENE—also called toluol—is used in solvent, paints, and gasoline. It is a central nervous system depressant and may cause headaches, fatigue, and confusion. There is evidence of embryo/*fetotoxin* in laboratory animals.

TOLUOL—*see* toluene.

TONKA BEAN is a natural bean. It is used in perfumes, snuff, pharmaceutical preparations, and deodorizer. Coumarin—also known as tonka bean camphor—is in the bean and is prohibited in food products. Coumarin is toxic by ingestion and is a *carcinogen.*

TOXALBUMIN—*see* albrin.

TOXAPHENE is a *pesticide* (insecticide). It is a *carcinogen* and bioaccumulates in the food chain.

TRIBROMOSALAN—*see* 3,4',5-tribromosalicyl-anilide.

3,4',5-TRIBROMOSALICYLANILIDE—also called tribromasalan—is used in soaps. It is not permitted in cosmetics. It is a *carcinogen.*

TRICALCIUM DIPHOSPHIDE—*see* calcium phosphide.

TRICHLOCARBAN (TCC)—also called 3,4,4'-trichlorocarbanilide—is a *pesticide* (bacterio-stat), disinfectant, and an antiseptic.

3,4,4'-TRICHLOROCARBANILIDE—*see* trichlocarban.

1,1,1-TRICHLOROETHANE—also called methyl-chloroform—is a *pesticide*, solvent, and cleaning agent. It is a central nervous system depressant.

TRICHLOROETHYLENE (TCE) is a *pesticide.* It is used in dry cleaning, degreaser, anaesthetic and analgetic, and in some correction paper fluids. It is a central nervous system depressant and a *carcinogen.*

TRIETHANOLAMINE (TEA) is a *pesticide*. It is used in cleaner, detergent, and cosmetics. It is an eye and skin *irritant*, and is a *carcinogen* in laboratory animals.

TRIISOPROPANOLAMINE is a *pesticide*, and is used in hair preparations. It is an eye and skin *irritant*.

TRYPTOPHAN—also called indole-alpha-aminoproponic acid—is an amino acid which is naturally occurring in some foods. It is a *carcinogen* and has resulted in death.

TUBATOXIN—*see* rotenone.

TURPENTINE—also called wood turpentine—is a *pesticide*, solvent, and paint thinner. It is a central nervous system depressant, skin and mucous membrane *irritant*, and questionable *carcinogen* in experiments.

TUTANE—*see* sec-butylamine.

U

UREA—also called carbamide—is a *pesticide*. It is used in animal feed, fertilizer, and cosmetics. It has experimental reproductive effects, and is a questionable *carcinogen* in experiments.

UREA PEROXIDE—*see* carbamide peroxide.

V

VANCIDE—*see* captan.

VANILLIN is used in *perfumes*, pharmaceuticals, and as a flavor. It is toxic by ingestion.

VAPONA—*see* dichlorovos.

VITAVAX—*see* carboxin.

VINYL CHLORIDE is used in the manufacture of polyvinyl chloride (PVC). It is a *carcinogen* and gives off toxic fumes if burned.

VINYL CYANIDE—*see* acrylonitrile.

VINYLFORMIC ACID—*see* acrylic acid.

VITAMIN B-5 CALCIUM SALT—*see* calcium pantothenate.

VITAMIN B-1—*see* methyl acrylate.

VITAMIN E has experimental reproductive effects.

VITAMIN H—*see* para-aminobenzoic acid.

VOLATILE OIL of MUSTARD—*see* allyl isothiocyanate.

W

WARFARIN is a *pesticide* and medical blood thinner. It is poisonous by ingestion and inhalation, has reproductive effects, causes fetal and physical abnormalities, and is an experimental *teratogen*.

WHITE ARSENIC—*see* arsenic trioxide.

WOOD TURPENTINE—*see* turpentine.

X

XANTHAN GUM is used in toothpaste, foods, and drilling fluids.

Z

ZINC ARSENATE is a *pesticide* and *wood preservative*.

ZINC ARSENITE is a *pesticide*.

ZINEB (Zinc ethylene bisdithiocarbamate) is a *pesticide* (fungicide). It is toxic by ingestion, a questionable *carcinogen*, and an experimental *teratogen*.

GLOSSARY

A

Abrasive—a chemical or product that is used to wear down something.

Absorbed/absorption—the ability of a chemical to be sorbed to soil and be released from soil with water or other liquid solvents (e.g., water in a sponge).

Adsorbed/adsorption—the ability for a chemical to sorb to soil and not be released with water and most other liquid solvents.

Active ingredients—an agent (i. e., a chemical, biological, etc.) that makes the product work.

Aggregate effect—the toxic effects that all the chemicals to which you could be exposed to would have on humans or the environment.

Alkaloid—an organic nitrogenous compound of vegetable origin.

Allergen—causes allergies.

Analgesic—a drug that relieves pain.

Anorexia—loss of appetite.

Anoxia—lack of oxygen; blood cannot carry adequate supply of oxygen.

Antagonistic effect—different reaction caused by opposite chemical forces.

Antibacterial—ingredient used as an antiseptic to kill or control microbes (e.g., bacteria).

Anticholinergic—antagonistic/opposing to the effects on nerve fibers.

Anthelmintic—agent which kills intestinal worms.

Antihypertensive agent—an agent that reduces blood pressure.

Antineoplastic—retards cellular growth.

Antioxidant—inhibits/slows down oxidation.

Antisporiatic—agent which kills spores.

Asphyxia—death caused by the lack of oxygen.

Assessment—as in risk assessment.

Astringent—used to constrict tissue.

B

Bioaccumulation—ability for a chemical to bioaccumulate/buildup in tissues, fat, etc.

Biometabolism—the process by which microbes break down or transform chemicals; also called biodegradation.

Break-down products products formed from the breaking down of a chemical. These products may or may not be more toxic than parent chemical—the starting chemical.

Bradycardia—a slow heartbeat.

C

Carcinogen—causes cancer.

Catalase—an enzyme in blood and tissues that contributes to the decomposition of hydrogen peroxide to form water and oxygen (a gas).

Caustic—a chemical that can burn, corrode, eat away, etc.

Central nervous system depressants (CNS-dep)—a group of chemicals (neurotoxics) that may cause dilated pupils, drowsiness, disorientation, vertigo, behavioral changes, nausea, vomiting, hypotension, and even symptoms of intoxication.

Chemosterilant—destroys the ability to reproduce.

Civet—a yellowish chemical secreted by the civet cat glands.

Coal tar pitch—is what is left after coal tar (from coal) is redistilled. Highly toxic by inhalation.

Co-biometabolism—a substance that would not ordinarily be biometabolized/biodegraded without an outside influence.

Colloidal—suspended particles in a media that does not immediately settle out of its suspension.

Combustible—a substance that will burn.

Co-metabolism—a substance that would not ordinarily be metabolized without an outside influence.

Contamination/ed—air, water, soil, plants and animals containing a chemical, microbe, virus, etc. that is not naturally occurring in that environment.

Counterirritant—aide to relieve skin irritation and itching. Often applied as ointment or lotion.

D

Dermatitis—inflammation of the skin.

Disintegrate—to break down; fall apart.

E

Edema—excess accumulation of fluids; accumulation of water; can cause swelling.

Electromagnetic field (EMF)—is radiation given off by electric forces/fields.

Emollient—softening and soothing effects.

Estrogenic—promotes sexual activity.

F

Fetotoxin—toxic to the fetus and can kill the fetus.

Fibrosis—tissue that is formed or in excess amounts.

Flammable—a substance that will easily ignite and burn.

Food additive—a chemical or substance that is allowed to be added to food.

Fumigant—a poisonous agent used on soil and in graineries.

Fungi—a group of plants including mushrooms, mold, yeast, etc.

Fungicide—*see* pesticide.

G

Glottis—part of the larynx.

H

Hazardous—a toxin or toxic chemical that is harmful to plants or animals including humans.

Hepatoxins—chemicals that effect the liver.

Histoplasmosis—a lung disease caused by a fungi (*Histoplasma capsulatum*) in fecal matter.

Hydrocarbon—an organic chemical consisting only of carbon and hydrogen elements.

Hypermotility—excessive motility of the stomach or intestine.

Hypomotility—abnormally slow motility of the stomach or intestine.

I

Inert ingredient—an agent (e.g., chemical, biological, etc.) which is in products but is not an active ingredient.

Intraperitoneal—by injection into the organism/body.

Intravenous—by injection into the vein.

Irritant—a substance that causes an irritation.

L

Lachrymator—a chemical that is an *irritant* to the eyes (e.g., tear gas).

Leaching—the ability of a chemical to move through soil with water or some other liquid solvent.

Lectin—a naturally occurring plant protein.

M

Metabolism—the process by which plants and animals break down or transform chemicals.

Methemoglobin—decrease of the oxygen carrying capacity of the blood.

Mutagen—causes mutations in living organisms.

N

Neurotoxin—causes brain damage.

O

Oncogen—an agent that causes non-malignant tumors.

Oxidation—combining with oxygen.

P

Pesticide—an ingredient used to mitigate/kill a pest. Includes all chemicals registered under the Federal

Insecticide Fungicide and Rodenticide Act (FIFRA) (i.e., herbicides, plant growth regulators, insecticides, fungicides, antifouling agents, etc.

Photo-degradation/transformation—the ability for a chemical to absorb sun light and break-down or build-up (to lower or heavier molecular weight molecules).

R

Reagent—a substance used because of its chemical and/or biological activity in a reaction, analysis, etc.

Residues—are what remain after completion of a chemical or physical process, such as evaporation, combustion, distillation, or filtration. Pesticide residues include any of the parent pesticide left in the environment as well as the break-down products or transformation products (a build-up product or a break-down product).

Risk assessment—a process to determine the potential risk to exposure to a toxin.

Run off—the ability of a chemical to move across soil with water or some other liquid solvent or with soil particles.

S

Seborrheaa—discharge from the sebaceous gland which forms an oily coating on the skin.

Subcutaneous—by injection under the skin; subcuteous tissue.

Sublime—the passage from one physical state to another (i.e., solid to vapor, ice to water vapor, etc.).

Synergistic—the effect/reaction that two or more substances cause together that they would not cause if separate.

T

Teratogen—causes birth defects/abnormalities from conception and birth.

Thromboembolism—the release of a blood clot that obstructs a blood vessel—may induce heart attacks.

TLV (threshold limit value)—is a standard concentration of airborne residues set by the

American Conference of Governmental Industrial Hygienists (ACGIH). These levels are a guide and are not to discern safety or other hazards.

Toxic—A poisonous substance that can cause illness or death even at low dosage.

Toxoplasmosis—a disease caused by microorganisms.

Transformation—the break down of a chemical to form a new chemical larger or smaller in molecular structure.

V

Vapor—a chemical in its volatile state/form (e.g., liquid water to water vapor).

Vesicant—agent that causes blisters.

Volatile/volatilization—the ability for a chemical to become a vapor and be released in the atmosphere.

W

Wood preservative—a *pesticide* that is used to protect wood from insects or microbes.

005

For a free catalog of Chatelaine Press publications: